Praise for *Green Wiz*

Using an integrated systems approach to ecologically sustainable living strategies, John Michael Greer tackles energy, economics, technology and organic production practices with approachable precision in *Green Wizardry*. Interested in a real "Green" revolution? Let *Green Wizardry* show you the way AND the why!

—Oscar H. Will iii, editor-in-chief, *GRIT* magazine
and author, *Plowing with Pigs and Other Creative,
Low-budget Homesteading Solutions*

Wanted: real wizards. Must be capable of guiding their communities through difficult times by exercising practical skills and applying down-to-earth knowledge of how things work. If you're interested in applying for the job, then read this marvelous book. Greer is the ideal guide; his wisdom and gentle humor animate every page. A magical career awaits you.

—Richard Heinberg, author, *The End of Growth*

Greer races back down the muddy track we've slogged up this past half century and shouts that he has found a turn we missed. It was clearly marked – flagged by economist Vince Taylor as "the Easy Path" and physicist Amory Lovins as "the Soft Path" – but we all just blew right by the signpost in our headlong rush up the hard and wide asphalt highway. We overestimated the difficulty of this road, and our skill at climbing, but that soft and easy path is still clearly marked, and open to us, if we just go back down a little ways. We actually have to. It is the only way forward.

—Albert Bates, founder, Global Village Institute
for Appropriate Technology and author,
The Post-Petroleum Survival Guide and *The Biochar Solution*

This is book is part nostalgic trip through what once could have been, and could have saved us, before we decided to go all in on unbounded economic growth and endless militarism, and part survival guide which, now that the results of this bout of folly are coming in, can allow families and communities to cut their burn rate in comfort and style. And it also lays out the foundations of a very useful and very necessary vocation to which all of us, young and old can aspire: Green Wizardry.

— DMITRY ORLOV, author of *Reinventing Collapse*
and *The Five Stages of Collapse*

Just as someone starting a business would get a book on business planning, this is the book for planning a greener life. Rather than being split into chapters, *Green Wizardry* is split into 36 lessons, each with an exercise at the end, where you are challenged to apply your new-found knowledge to your own life or delve more deeply into a subject. If there were a class on green living, this would be the ideal text!

— DEBORAH NIEMANN, author of *Homegrown and Handmade,*
Ecothrifty, and *Raising Goats Naturally*

Green Wizardry

Conservation, Solar Power, Organic Gardening, *and other* Hands-On Skills *from the* Appropriate Tech Toolkit

John Michael Greer

new society
PUBLISHERS

Cover design by Diane McIntosh. Cover image © Rose Ann Janzen

Printed in Canada. Second printing September 2015.

Paperback ISBN: 978-0-86571-747-3
eISBN: 978-1-55092-539-5

Inquiries regarding requests to reprint all or part of *Green Wizardry*
should be addressed to New Society Publishers at the address below.

To order directly from the publishers, please call toll-free (North America)
1-800-567-6772, or order online at www.newsociety.com

Any other inquiries can be directed by mail to:

New Society Publishers
P.O. Box 189, Gabriola Island, BC V0R 1X0, Canada
(250) 247-9737

New Society Publishers' mission is to publish books that contribute in fundamental
ways to building an ecologically sustainable and just society, and to do so with the
least possible impact on the environment, in a manner that models this vision. We
are committed to doing this not just through education, but through action. The
interior pages of our bound books are printed on Forest Stewardship Council®-
registered acid-free paper that is **100% post-consumer recycled** (100% old growth
forest-free), processed chlorine free, and printed with vegetable-based, low-VOC
inks, with covers produced using FSC®-registered stock. New Society also works to
reduce its carbon footprint, and purchases carbon offsets based on an annual audit
to ensure a carbon neutral footprint. For further information, or to browse our
full list of books and purchase securely, visit our website at: www.newsociety.com

LIBRARY AND ARCHIVES CANADA CATALOGUING IN PUBLICATION

Greer, John Michael, author

Green wizardry : conservation, solar power, organic gardening, and other hands-on
skills from the appropriate tech toolkit / John Michael Greer.

Includes bibliographical references and index.

ISBN 978-0-86571-747-3 (pbk.)

1. Sustainable living. 2. Energy conservation. 3. Organic gardening.
4. Vegetable gardening. 5. Green technology. 6. Dwellings—Energy conservation. I.
Title.

GE196.G74 2013   640.28'6   C2013-902582-0

Contents

A Time for Wizardry

A little more than three decades ago, as a geeky kid from an ordinarily troubled American family, I began my freshman year of college at a midsized state university. I had, as eighteen-year-olds usually have, a laundry list of things I wanted to do with my life, spread out along the spectrum from the plausible to the preposterous. Novels by J.R.R. Tolkien and Hermann Hesse, books of environmental philosophy by Theodore Roszak and E. F. Schumacher, and a well-thumbed copy of the *Tao Te Ching* were the intellectual compasses by which I hoped to find my way. At the time, I was considering a career in what was then the booming field of appropriate technology.

Those of my readers who were not around in the 1970s may never have encountered the phrase "appropriate technology," and plenty of people who were around then seem to have done their level best to forget it. Still, in a time of energy crises and gas lines, when a great many people all over the industrial world recognized the absurdity of pursuing limitless growth on a finite planet, it was a label that anchored a great many hopeful visions of the future.

E. F. Schumacher, the maverick economist whose book *Small is Beautiful* was one of my guiding lights in those days, launched the appropriate technology movement back in the 1950s in response to the needs and possibilities of the nonindustrialized world. He called it "intermediate technology"—intermediate, that is, between the traditional folk technologies of the global South and the energy- and capital-intensive technologies of the industrial nations. As the energy crises of the 1970s made it clear that those same energy- and capital-intensive technologies were no more sustainable in rich nations than in poor ones, a generation of green innovators adapted the phrase, and the concept, to the demands of the approaching Age of Limits.

In the world of the twenty-first century, appropriate-tech mavens argued, the cheap abundant energy and resources that supported the extravagant machinery of twentieth-century industrial nations would inevitably run short. Before that happened, a new breed of technology had to be invented and put into production. The new technologies they hoped to pioneer would use energy and resources sparingly; they would work with the cycles of nature rather than against them; they would meet human needs without placing unsustainable burdens on the biosphere. All over the world in those days, you could find little nonprofits on shoestring budgets and small companies run by basement entrepreneurs, hard at work making that dream a reality.

To a remarkable extent, they succeeded. This is the fact that most often gets left out of the story on those rare occasions when the appropriate-tech project is remembered at all. By the time I started college in 1980, the core technologies had been invented, the most challenging problems solved. In dozens of different ways, varying with local conditions and resources, people in the appropriate tech movement proved that it was possible to live sustainably, and even comfortably, on a small fraction of the energy and resources most middle-class people in the industrial world thought they needed to sustain their lives. As the new technologies moved toward maturity, a great many of us recognized that with hard work and a willingness to make modest sacrifices in order to build a better world for our grandchildren, the world could manage a relatively smooth transition into the Age of Limits.

That wasn't what happened. Instead, the world's industrial nations threw away the promise of a relatively easy transition into the future and committed themselves to a trajectory that guaranteed they would face a head-on collision with the hard limits of a finite planet. Here in the United States, the tipping point was Ronald Reagan's election as president, which gave a radical faction within the Republican party the chance to dismantle the environmental policies and green initiatives of the previous decade. The same thing happened elsewhere in the world about the same time, as "conservative" parties (the word belongs in quotes, since buying short-term prosperity at the price of long-term calamity is not a conservative policy in any meaningful sense of the term) discovered that they could win elections and reshape the political dialogue by insisting that the world could keep using up irreplaceable resources at unsustainable rates and never have to suffer the consequences.

The conservatives' claim never rested on anything more solid than partisan rhetoric, but it allowed millions of people to justify their abandonment of the steps toward sustainability that had been taken in the previous decade, and it also provided an excuse for pumping newly drilled oil fields in the North Sea and the Alaskan North Slope as fast as geology would permit, flooding world markets with oil and sending the price plummeting to the neighborhood of US$10 a barrel—its lowest level, corrected for inflation, in all of human history. The flood of cheap oil launched a lavish if temporary economic boom across the industrial world, sweeping aside the nascent culture of conservation just as it was beginning to bear fruit.

There were some people, to be sure, who stayed with the appropriate-technology project even after the Reagan era's temporary glut of cheap energy made it unfashionable. I was one of them. I left college without a degree when it became clear that my hopes of making a career in appropriate tech were headed nowhere; even so, I was awarded a Master Conserver certificate from the Washington State Energy Extension Program in 1985 (just before it lost its funding) and I kept growing organic gardens and conserving energy and resources straight through the years of extravagance that followed. I wasn't the only person who held onto the dream, though it sometimes felt that way. Still, our ideas were often treated as an embarrassment, even among those people who claimed they were concerned about the environment and the future of the planet.

The irony of that response grew even more pointed as the twenty-first century neared, and the Age of Limits began to show up on schedule. The global peak of petroleum production found a new name—Peak Oil—and began its long journey from the fringes to the front pages of the daily paper; hard limits to other resources, ranging from the phosphorus feedstocks for industrial agriculture to the water that makes it possible for crops to grow at all, came into sight; climate change driven by industrial society's unhelpful habit of treating the atmosphere as an aerial sewer stopped being a theoretical issue and started sending the global climate spinning out of control worldwide. The list could go on a great deal further, but I trust my readers will have gotten the point, which is that the concerns that drove the appropriate tech movement in the 1970s are deeply relevant to our lives today and the easy dismissal of those concerns by the pseudo-conservatives of the 1980s and 1990s were hopelessly naive.

Yet this belated discovery didn't send people back, as it should have, to the practical discoveries and philosophical insights that emerged from the last sustained effort to deal with the predicament we're now facing. Quite the contrary: most of the generation that turned its back on the appropriate tech movement at the time of the Reagan counterrevolution still gets visibly uncomfortable if the subject gets brought up in their presence, and the subject of appropriate tech has been pushed so far down the hole where our culture keeps its unwelcome memories that the generations coming of age since then don't know the least thing about it. Instead, the one kind of recycling that caught on in the Reagan years—the recycling of failed ideological responses to the ecological crisis of industrial civilization—has become standard practice at all points along the political spectrum. As each year passes and the Age of Limits closes in, one party talks about the free market and all the oil that would be out there if only government regulation would get out of the way. The other party talks about marvelous new innovations that never quite go into production, or they propose grand radical agendas that only attract the usual activists. Both sides try not to notice that none of these things is doing anything to divert the trajectory of our age from a messy end.

Many people today know this, or at least sense it. Especially, but not only, in North America, an awareness is spreading through the crawlspaces of our culture that the current round of troubles isn't just a speedbump on the road to the shiny future our society's myths promise us. The sense that something has gone desperately wrong, right down at the core of the world we've created for ourselves, forms the background to most of our culture these days. The popularity of soothing narratives about the future is a marker for just how pervasive that background has become; it's only when fears are inescapable that efforts at mass reassurance find a market.

Still, those reassurances are becoming very thin these days. These days, when I give a talk on the future, I begin it this way: "Remember all those scientists a few decades back who warned that if we didn't make drastic changes in the very near future, we'd be in deep trouble in the early twenty-first century? Well, guess what." Much more often than not, nobody argues. Some of them, to judge by the comments and questions I field, are already trying to fit their minds around a future in which we will all have to get by on much less energy and fewer resources than we use today, but many more simply hope they'll be lucky enough to die before it hits—I've had people, not all of them elderly, express this latter hope to me in so many words.

Even among those who insist that everything will work out— those who claim that the center of the Earth really is full of limitless oceans of oil, or that we can all move into lifeboat ecovillages and embrace an abundant new lifestyle, or that the Rapture, the Singularity, or some other *deus ex machina* will save them from the future—I have to question how many people actually believe what they claim to believe. I know far too many people who insist the world will end soon, for example, who are still putting money into their retirement accounts, and it's been an open secret for the last decade that the vast majority of people who like to imagine living in a lifeboat ecovillage haven't been willing to lift a finger or spend a dime to bring such a project into being.

I'd like to suggest that this mismatch between what people claim to believe and what they believe enough to put into action arises because daydreams about lifeboat ecovillages and planet-ending catastrophes share a common purpose, one that's also shared by the assortment of unproven but heavily-promoted technologies that claim to be able to solve our energy crisis without requiring anyone to give up the extravagant energy and resource-wasting lifestyles that pass for normal in the industrial world these days. That common purpose has nothing to do with addressing the crises these things claim to address; instead, it's a matter of giving people something more pleasant to think about than the future that's breathing down our necks. Even planet-ending catastrophes, after all, are more pleasant to think about than the couple of centuries of ragged decline, impoverishment, and population loss that define the usual fate of a failed civilization. If we all get blown to kingdom come in one vast fireball, at least it's over quickly, and we can get some consolation in the few seconds before we're vaporized by telling ourselves that it wasn't our fault.

That state of denial is among the dominant forces in the collective psychology of the industrial world today. Its roots go down into some of the most tangled and murky corners of the modern mind. The issue that I want to explore here is whether there are options for constructive action at a time when most people think that hiding under the mattress is a useful response to the near future. I believe there are such options, and one of them may best be understood by borrowing a metaphor from history.

A few years ago, as part of the other side of my career as a writer,

I helped translate a very strange book from the Middle Ages. Its title is *Picatrix*, and it's one of the very few surviving examples of that absolute rarity of medieval literature, a textbook for apprentice wizards.[1] Those of my readers who grew up on stories about Merlin, Gandalf et al. may be startled to learn that those characters, legendary or fictional as they are, were modeled on an actual profession that flourished in the early Middle Ages and remained common until the bottom fell out of the market at the end of the Renaissance.

By "wizard," I don't mean your common or garden variety fortune teller or ritual practitioner; we have those in abundance today. The wizard of the early Middle Ages was a freelance intellectual whose main stock in trade was good advice, though that came well frosted with incantation and prophecy as needed. He had a working knowledge of astrology, which filled the same role in medieval thought that physics does today, and an equally solid knowledge of ritual magic, but his training didn't begin or end there. According to *Picatrix*, a wizard needed to have a thorough education in agriculture; navigation; political science; military science; grammar, languages, and rhetoric; commerce; all the mathematics known at the time, including arithmetic, geometry, music theory, and astronomy; logic; medicine, including herbal medicines and poisons; the natural sciences, including meteorology, mineralogy, botany, and zoology; and metaphysics—in effect, the sum total of scientific learning that had survived from the classical world.

This might not sound like the sort of education you would expect to get at Hogwarts, the school for young wizards in J. K. Rowling's *Harry Potter* novels, and that's exactly my point. Whether you believe that the movements of the planets foretell events on Earth, as almost everyone did in the Middle Ages, or whether you think astrology is simply a clever anticipation of game theory that gets its results by inserting random factors into strategic decisions to make them unpredictable, you'll likely recognize that a soothsayer with the sort of background I've just sketched out would be well prepared to offer sound advice on most of the questions that might have perplexed a medieval peasant, merchant, baron, or king. Equally, of course, someone so trained would hardly be restricted in his choice of active measures to incantations alone. This is arguably why so many medieval kings and barons had professional sorcerers and soothsayers on staff despite the fulminations of all the major religions of the age. It also explains why wizards less adept at social climbing found a bumper crop of customers lower down the social ladder.

The origins of this profession are, if anything, even more interesting. A useful study by historian Pierre Riché has explored in detail how the educational system of the late Roman world imploded as its economic and social support systems crumpled beneath it.[2] In Europe (matters were a little more complex in the Muslim world), that system was replaced by a monastic system of education that, in its early days, fixated almost entirely on scriptural and theological studies. Methods of training young aristocrats fixated even more tightly on the skills of warfare and government. Only among families that held onto a tradition of classical learning did something like the old curriculum stay in use. Riché notes that while this custom continued, those who learned philosophy, one of the core studies in that curriculum, were widely suspected of dabbling in magic. It's not too hard to connect the dots and see how a subculture of freelance intellectuals, equipped with unusual knowledge and a willingness to wander well outside the boundaries set by the culture of their time, could have emerged from that context.

All this may seem worlds away from the issues raised a few paragraphs ago, but there's a direct connection. The wizards of the early Middle Ages were individuals who recognized the value of certain branches of knowledge that were profoundly unpopular in their time, and they took it on themselves to preserve the knowledge and make connections with those who shared their interests—or at least wanted to make practical use of the skills that the knowledge made possible. No government support was available for the transmission of classical knowledge in the sixth and seventh centuries CE, and the popular movements of the time—when they weren't simply stampeding mobs trying to get out of the way of the latest round of barbarian invasions—were no more likely to help. How much of a role wizards might have played in the transmission of classical learning to the future is anyone's guess; records of their activities are very sparse. It's clear, however, that they were an intellectual resource much used during an age when few other resources of the kind were available.

I've come to think that something like the wizards of the early Middle Ages, focused on a somewhat different body of skills, may be one of the best options we have available today. Certain branches of practical knowledge, thoroughly learned and thoroughly practiced by a relatively modest number of people, could be deployed in a hurry to help mitigate the impact of the energy shortages, economic dislocations, and systems breakdowns that await us in the years ahead. I'm sure my readers have their own ideas about the kind of knowledge that might be best suited to that challenge, but the appropriate technology movement of the 1970s makes a particularly good focus for such a project.

That movement, as already suggested, differed in important ways from the efforts toward sustainability that get most of the media coverage today. The kind of high-cost, high-profile eco-chic projects that receive the lion's share of attention were of little interest back then. Instead, effort focused on simple technologies that could be put to work by ordinary people—people without six-figure incomes, doing the work themselves, and using readily available tools and resources. Most of the technologies were evolved by basement-shop craftspeople and small nonprofits working on shoestring budgets. These technologies were field-tested by thousands of people, many of whom built their own backyard versions and then wrote about the results in the letters column of *Mother Earth News*.

The toolkit that evolved out of this process was a remarkably well-integrated, effective, and cost-effective set of approaches that individuals, families, and communities can use to sharply reduce their dependence on fossil fuels and the industrial system. It was not, I should probably point out, particularly esthetic, unless you happen to like a lively fusion of down home funk, late twentieth-century garage-workshop modern, and hand-dyed back-to-the-land hippie paisley. Those of my readers who own houses and are still fretting about resale value and haven't yet figured out that this figure will be denominated in imaginary numbers for the next several decades at least, will likely run screaming from it. Those who were incautious enough to buy homes in suburban developments with restrictive covenants will have to step carefully, at least until their neighbors panic. Apartment dwellers will have to pick and choose a bit; on the other hand, those of my readers who will spend time living in tarpaper shacks before the current Great Recession ends—and I suspect that many people will have that experience, as many people did the last time the economy lost touch with reality and imploded—will find that very nearly everything the appropriate-tech people did will be well within their reach.

What's included in the package we're going to explore? Intensive organic gardening, with its support technologies of composting, season extenders, and low-tech food preservation and storage methods; small-scale chicken and rabbit raising, and home aquaculture of fish; simple solar greenhouses, which make the transition from food to energy by providing heat for homes as well as food for the table; other retrofitted passive solar heating technologies; solar water heating; a baker's dozen or more methods for conserving hot or cool air with little or no energy input; and a good deal more along the same lines, all of it geared toward backyard gardens, basement workshops, easily available resources, and modest budgets. None of these things, it's only fair to say, will save the world, if that hackneyed phrase means maintaining business as usual on some supposedly sustainable basis. What the toolkit can do is make human life a good deal less traumatic and more livable in an age of serious energy shortages and economic troubles.

Given that serious energy shortages and economic troubles are everyday phenomena in much of the world right now, and are far from unheard of right here in the industrial world, it's bizarre that the alternative approaches I've just listed remain entirely off our collective radar screen. Among the things that make it difficult for many people to factor these alternatives into their future is that they aren't, so to speak, plug-and-play components for existing lifestyles; they presuppose radically different relationships among land, resources, farmers, crops, and consumers. As they expand into the spaces left blank by today's faltering industrial society, the new social forms defined by these relationships differ so starkly from existing ways of doing things that many people have trouble making sense of them.

This same pattern pervades nearly all current debates about the future. Consider the endless bickering over the potential of renewable energy in the media and the internet. Most of that bickering assumes that the only way a society can or should use energy is the way today's industrial nations use energy. Thus you see one side insisting that windpower, say, can provide the same sort of instantly accessible and abundant energy supply we're used to having, using some equivalent of the same distribution systems and technologies we're used to using, while the other side—generally with better evidence—insists that it cannot.

What inevitably gets missed in these debates is the fact that it's entirely possible to have a technologically advanced and humane society without having electricity on demand from sockets on every wall across the length and breadth of a continent, or mortgaging our future to allow individuals to zoom around in hopelessly inefficient personal vehicles on an extravagant system of highways. The sooner we start thinking about what kinds and forms of energy wind turbines are actually best suited to produce, rather than trying to force them onto the Procrustean bed of an electrical grid that was designed to exploit the very idiosyncratic kinds of energy we get from fossil fuel supplies, the sooner windpower can be put to use building an energy system for the future—instead of being wasted in a futile attempt to prop up the obsolete one we've inherited from the recent past. What stands in the way of this recognition is the emotional power of today's ideology of progress, with its implicit assumption that the way we happen to do things must be the best, or even the only, possible way to do them.

Seeing past that assumption was among the core achievements of the backyard gardeners and basement inventors of the appropriate tech movement of the 1970s. Their capacity to glimpse other possibilities gave the toolkit they evolved a creative edge that too many current proposals lack. For this reason, among others, there's a point to revisiting the skills of a time in which the limits to growth were recognized as an imminent reality. It's even possible that those who are willing to do so, and learn these unfashionable but potentially vital skills, might follow the very old example mentioned earlier, and become the green wizards of the age to come.

For that, I have come to think, is one of the things that a world on the brink of a long descent into a deindustrial age most needs: green wizards. By this, I mean individuals who embrace the task of learning, practicing, and thoroughly mastering a set of mostly forgotten skills to use and to share with others. This isn't a subject where armchair theorizing counts for much—as every wizard's apprentice learns, what you really know is measured by what you've actually done—and it's not going to earn anyone a living any time soon, either—though it can help almost anyone make whatever living they earn go much further. Nor, again, will it prevent the unraveling of the industrial age and the coming of a new era of hard material limits. What it can do, if enough people seize the opportunity, is make the road to that new era much more bearable than it will otherwise be.

This same principle can be applied to many things besides the suite of green technologies I studied as a budding appropriate-tech geek during the late 1970s and early 1980s. Still, a set of skills evolved during the last round of energy crises and economic contraction is likely to be particularly useful in this round. It seems to me, however, that there are deeper lessons to be learned in the hard contrast between the eager hopes of the appropriate-tech movement I studied as a young man, and the brittle cynicism that clamped down over much of the industrial world in the decade that followed. A look at those deeper lessons belongs to a later section of this book, but it may be helpful for readers to keep the contrast in mind as we glance back across the decades to a time when it seemed as though we might actually find the courage to make a better future for ourselves and our world.

PART ONE

Principles

ᨆ LESSON 1 ᨆ
Introducing Energy

It's not accidental that the appropriate tech movement of the 1970s was brought into being by the experience of energy crisis, or that tools and insights having to do with energy played a central role in that movement. In the most pragmatic of senses, understand energy and you understand the whole art of appropriate technology. In the broadest of senses, understand energy and you understand the predicament that is looming up like a wave in front of the world's industrial societies, and you also understand what we can and cannot expect to get done in the relatively short time we have left before the pressures unleashed by that predicament crest, break, and wash most of the modern world's certainties away.

Important as it is, though, energy doesn't stand alone. Two other concepts join the concept of energy to provide the central triad of principles that undergird this book and the perspectives and practices it explores. The first of these additional concepts is matter; the second is information. These three—energy, matter, and information—flow constantly through every whole system, in nature or in human society. Understand these flows and you understand the system. Each of the three, though, follows its own rules, so we'll explore them one at a time.

We can start with some basic definitions. Energy is the capacity to do work. It cannot be created or destroyed, but the amount and kind of work it can do can change. The more concentrated it is, compared to its surroundings, the more work it can do; the less the difference in its concentration and the background level of energy around it, the less work it can do. Left to itself, it moves from more concentrated to more diffuse forms over time, so everything you do with energy has a price tag, measured in a loss of concentration. These are the groundrules of thermodynamics, and everything a green wizard does comes back to them in one way or another.[3]

Some examples will help show how these rules work. In energy terms, for instance, a garden bed is a device for collecting solar energy by way of the biochemical dance of photosynthesis. Follow a ray of sunlight from the seething thermonuclear cauldron of the sun, across 93 million miles of hard vacuum and a few dozen miles of atmosphere, until it falls on the garden bed. About half the sunlight reflects off the plants, which is why the leaves look bright green to you instead of flat black; most of the rest is used by the plants to draw water up from the ground and expel it as water vapor into the air; a few percent is caught by chloroplasts—tiny green disks inside the cells of every green plant—and used to turn water and carbon dioxide into sugars, which are rich in chemical energy and power the complex cascade of processes we call life.

Most of those sugars are used up keeping the plant alive. The rest are stored for the plant's future needs, though a percentage of them get hijacked if some animal eats the plant. Most of the energy in the plants the animal eats gets used up keeping the animal alive; the rest get stored until another animal eats the first animal, and the process repeats. Sooner or later, an animal manages to die without immediately ending up in something else's stomach, and its body becomes a lunch counter for all the creatures—and there are a lot of them—that make a living by cleaning up dead things. By the time they're finished with their work, the last of the energy from the original beam of sunlight that fell on the garden bed has been lost to the food chain.

What happens to it then? It turns into diffuse background heat. That's the elephant's graveyard of thermodynamics, the place useful energy goes to die. When you do anything with energy— concentrate it, move it, change its form—a price has to be paid in diffuse heat. All along the chain from the sunlight first hitting the leaf to the last bacterium munching on the last scrap of dead fox, what isn't passed onward is turned directly or indirectly into heat so diffuse that it can't be made to do any work other than jiggling molecules a little. The metabolism of the plant generates a trickle of heat; the friction of the beetle's legs on the leaf generates a tiny pulse of heat; the mouse, the snake, and the fox all turn most of the energy they take in into heat, and all that heat radiates out into the great outdoors, warming the atmosphere by a tiny fraction of a degree, and slowly spreading up and out into the ultimate heat sink of deep space.

That's one example. For another, let's take a solar water heater, the simple kind that's basically a tank in a glassed-in enclosure set on top of somebody's roof. Once again we start with a ray of sunlight crossing deep space and Earth's atmosphere to get to its target. The light passes through the glass and slams into the black metal of the water tank, giving up much of its energy in the form of heat. Inside the metal is water, maybe fifty gallons of it; it takes a fair amount of heat to bring fifty gallons of water to the temperature of a good hot bath, but the steady pounding of photons against the black metal tank will do the trick in just a few hours.

Most of what makes building a solar water heater complex is a matter of keeping that relatively concentrated heat in the water where it belongs, instead of letting it leak out as—you guessed it— diffuse background heat. The glass in front of the tank is there to keep moving air from carrying heat away, and it also helps hold heat in by way of a clever bit of physics: most of the energy that matter absorbs from visible light downshifts to infrared light as it tries to escape, and glass lets visible light pass through it but reflects infrared back the way it came. (This is known as the greenhouse effect, by the way, and we'll be using it later on in this book, not least in the context of actual greenhouses.) All the surfaces of the tank that aren't facing the sun are surrounded by insulation to keep heat from sneaking away, and the pipes that carry hot water down from the heater to the bathtub and other uses are wrapped with insulation. Even so, some of the energy slips out from the tank, some of it makes a break for it through the insulation around the pipes, and the rest of it starts becoming background heat the moment it leaves the faucet.

There are five points I'd like you to take home from these examples. The first is that both the plant and the solar heater function as a result of the same process: the flow of energy from the sun to Earth. Start looking at everything that goes on around you as an energy flow that starts from a concentrated source—almost always the sun—and ends in diffuse heat. If you do this, you'll find that a great deal of the material in this book is simply common sense—and a great many of the habits that are treated as normal behavior in our society will suddenly reveal themselves as stark staring lunacy.

The second point to take home is that natural systems, having had much more time to work the bugs out, are much better at containing and using energy than most human systems are. The solar water heater and the house with its natural gas furnace take concentrated energy, put it to one use, and then lose it to diffuse heat. A natural ecosystem, by contrast, can play hot potato with its own input of concentrated energy for a much more extended period, tossing it from one organism to another for quite a while before all of the energy finally follows its bliss. The lesson here is simple: by using existing natural systems to do things, green wizards can take advantage of two billion years of evolution, and by paying close attention to the ways that natural systems do things, green wizards can get hints that will make human systems less wasteful.

The third point is that energy doesn't move in circles. In the next lesson, we'll be talking about material substances, which do follow circular paths; in fact, they do this whether we want them to do so or not, which is why the toxic waste we dump into the environment ends up circling back around into our food and water supply. Energy, though, follows a trajectory with a beginning and an end. The beginning is always a concentrated source, which again is almost always the sun; the end is diffuse heat. Conceptually, you can think of energy as moving in straight lines, cutting across the circles of matter and the far more complex patterns of information gain and loss. Once a given amount of energy has followed its trajectory to the endpoint, for all practical purposes, it's gone; it still exists, but the only work it's capable of doing is making molecules vibrate at whatever the ambient temperature happens to be.

The fourth point is that while energy is the capacity to do work, it can't do work in a vacuum. To make energy do whatever work you have in mind for it—whether that work consists of growing plants, heating water, or anything else—matter, information, and additional energy have to be invested. The plant needs carbon dioxide, water, and an assortment of minerals, as well as the information in its DNA and a stock of energy stored up in sugars from previous sunlight, in order to turn each new photon into useful energy. The solar water heater has equivalent requirements. Far more often than not, these secondary requirements impose limits that are far stricter than the limits imposed by the energy flow itself.

The fifth and final point, which follows from the third and fourth, is that for practical purposes, energy is finite. It's common for people these days to insist that energy is infinite, with the implication that human beings can walk off with as much of it as they wish. This is an appealing fantasy, flattering to our collective ego, and it plays a central role in backing our culture's myths of perpetual technological progress and limitless economic growth.

As ecologist Garrett Hardin pointed out quite a while ago, though, it's also nonsense. In his useful book *Filters Against Folly* (which should be required reading for any student of green wizardry), Hardin showed that words such as "infinite," "limitless," and "boundless" are thoughtstoppers rather than useful concepts because the human mind can't actually think about infinity.[4] When people say "X is infinite," what they are actually saying is "I refuse to think about X."

Still, there's a more specific sense in which talk about infinite energy is nonsense. At any given place and time, the amount of energy that is available in a concentration and a form capable of doing any particular kind of work is finite, often distressingly so. Every ecosystem on Earth has evolved to make the most of whatever energy is available, whether that energy takes the form of equatorial sunlight shining down on the Amazon rain forest, chemical energy in sulfur-laden water surging up from hot springs at the bottom of the sea, or fat stored up during the brief Arctic warm season in the bodies of the caribou that attract the attention of a hungry wolf pack.

Thus it's crucial to recognize that useful energy is always limited, and it usually needs to be coaxed into doing as much work as you want to get done before it gets away from you and turns into diffuse background heat. This is true of any whole system, a garden as much as a solar hot water system, a well-insulated house, or any other project belonging to the field of appropriate tech. Learn to think in these terms and you're well on your way to becoming a green wizard.

Exercise for Lesson 1

Draw a rough flow chart for the flow of energy through some part of your life. Take a piece of paper, draw a picture of the sun at the top, and draw a trash can at the bottom; label the trash can "Background Heat." Now draw the important components in any system you want to understand, and draw arrows connecting them to show how the energy moves from one component to another. If you're sketching a natural system, draw in the plants, the herbivores, the carnivores, and the decomposers, and sketch in how energy passes from one to another, and from each of them to the trash can; if you're sketching a human system, the energy source, the machine that turns the energy into a useful form, and the places where the energy goes all need to be marked in and connected.

Do this with a variety of different energy flows, and do at least a little research on the internet or at your local public library to check your facts. It doesn't matter at this stage if you get all the details right; the important thing is to start thinking of energy in terms of finite flows, and get past the fantasy of limitlessness.

LESSON 2
Introducing Matter

Of the three factors that flow through every whole system—energy, matter, and information—matter is the messiest. If you have small children in your house, you have an advantage in making sense of matter, because you already know the most important of the rules that govern matter: it does not behave itself. It does not do what it's told. As you found out around the age of two, to your ineffable delight and your mother's weary annoyance, matter gets all over everything, especially when stomped. Most people discover this in childhood and then spend the rest of their lives trying to forget it. One of the ways they forget it in modern industrial cultures is by pretending that matter acts like energy.

Get a piece of paper and a pen, and I can show you how that works. At the top of the paper, draw a picture of Santa Claus in his sleigh, surrounded by an enormous pile of gifts. Label it "infinite material resources." In the middle, draw a picture of yourself sitting on heaps of consumer goodies; put in some twinkle dust, too, because we'll pretend (as modern industrial societies do) that the goodies somehow got there without anybody having to work sixteen-hour days in a Third World sweatshop to produce them. Down at the bottom of the paper, draw some really exotic architecture, and put a Chamber of Commerce sign out in front that says "Welcome to Away." You know, Away—the mysterious place where no one has ever been, but where stuff goes when you don't want it around anymore. Now draw one arrow going from Santa to you, and another from you to Away.

Does this picture look familiar? It should. It has the same pattern as a very simple energy flow diagram, just like the one you sketched in the previous lesson. Here, Santa is the energy source and Away is the diffuse background heat that all energy turns into sooner or later. Although this sort of diagram works perfectly well with energy, it doesn't work worth beans with any material substance. Nevertheless, this is how people in modern industrial societies are taught to think about matter.

As an antidote to that habit of thinking, take your Santa diagram, crumple it up with extreme prejudice, and throw it across the room. It would be particularly helpful if Fido is in the room with you, decides you've thrown a ball for him to chase, and he comes trotting eagerly back to you with the diagram in his mouth, having reduced it to a drool-soaked mess. At that moment, as you meet Fido's trusting gaze and try to decide whether it's more bother to go get a real ball for him to play with or take the oozing object that was once your drawing and then wipe a couple of tablespoons of dog slobber off your hand, you will have learned one of the great secrets of green wizardry: matter moves in circles, especially when you don't want it to do so.

Back in my schooldays, corporate flacks trying to head off the rising tide of popular unhappiness with what was being done to the American environment had a neat little slogan: "The solution to pollution is dilution." They were dead wrong, and because their slogan got put into practice far too often, a sizable number of people and a much greater number of other living things ended up just plain dead. Dilute an environmental toxin all you want, and it's a safe bet that a food chain somewhere will concentrate it right back up for you and serve it on your plate for breakfast. It's hard to think of anything more dilute than the strontium-90 dust that was blasted into the upper atmosphere by nuclear testing and scattered around the globe by high-level winds; that didn't keep it from building up to dangerous levels in cow's milk, and shortly thereafter, in children's bones.

A similar difficulty afflicts the delusion that we can put something completely outside the biosphere and make it stay there. Proponents of nuclear power who don't simply dodge the issue of radioactive waste altogether treat that issue as a minor matter. It's not a minor matter; it's the most critical of half a dozen disastrous flaws in the shopworn 1950s-era fantasy of limitless nuclear power still being retailed by the nuclear industry's few remaining cheerleaders. A nuclear fission reactor—any nuclear fission reactor—produces wastes so lethal they have to be isolated from the rest of existence for a quarter of a million years—in other words, fifty times as long as all of recorded history. In theory, containing high-level nuclear waste is possible; in practice, Murphy's law is the safer guide. In the real world, it's certain that sooner or later, things go wrong.

Despite the best intentions and the most optimistic rhetoric, in a hundred years, or a thousand, or ten thousand, by accident or malice or the sheer cussedness of nature, that waste is going to leak into the biosphere, and once that happens, anyone and anything that comes into contact with even a few milligrams of it will suffer a miserable death. The more nuclear power we generate now, the more of this ghastly gift we'll be stockpiling for the people of the far future. If one of the basic concepts of morality is that each of us ought to leave the world a better place for those who come after us, there must be some sort of gold medal for selfish malignity in store for the notion that, to keep our current civilization going a little longer, we're justified in making life shorter and more miserable for people whose great-great-great-grandparents haven't even been born yet.

This extreme case illustrates a basic rule of green wizardry: *There is no such place as Away.* You can throw matter out the front door all you want, but it will inevitably circle around while you're not looking and come trotting up the back stairs. There's a great deal of pop mysticism these days about how wonderful it is that everything in the universe moves in circles; it's true enough that matter moves in circles, though energy and information don't, but it's not always wonderful. If you recognize matter's habits and work with them, you can get it to do some impressive things as it follows its rounds, but if you aren't watching it closely, it can just as easily sneak up behind you and clobber you.

The trick to making matter circle in a way that's helpful to you is twofold. The first half is figuring out every possible way it might circle; the second is to make sure that as it follows each of those pathways, it goes through transformations significant enough to make it harmless. I hope I won't offend anyone's delicate sensibilities by using human feces as an example. The way we handle our feces in most American communities is frankly bizarre; we defecate in clean drinking water, for heaven's sake, and then flush it down a pipe without the least thought of where it's going. Where it's going, most of the time, is into a river, a lake, or the ocean, and even after sewage treatment, you can be sure that most of what's in your bowel movements is going to land in the biosphere pretty much unchanged because mushing feces up in water and then dumping some chlorine into the resulting mess doesn't change them enough to make a difference.

Consider the alternative of a composting toilet and a backyard garden. Instead of dumping feces into drinking water, you feed them to hungry thermophilic bacteria. When the bacteria get through with the result, you put the compost into the middle of your main compost pile, where it feeds a more diverse ecosystem of microbes, worms, insects, fungi, and the like. When those organisms are done with it, you dig the completely transformed compost into your garden, and soil organisms and the roots of your garden plants have at it. When you pick an ear of corn from your garden, some of the nutrients in the corn got there by way of your toilet, but you don't have to worry about that. The pathogenic bacteria that make feces dangerous to human beings, having grown up in the cozily sheltered setting of your bowels, don't survive long in the Darwinian environment of a composting toilet, and any last stragglers get mopped up in the even more ruthless ecosystem of the compost pile.

In the same way, the inedible parts of garden vegetables can be put into the compost pile or, better still, fed to chickens or rabbits, whose feces can be added to the compost pile, so that plant parasites and diseases have less opportunity to ride the cycle back to the plants in the garden. You can cycle other parts of your household waste stream into the same cycle; alternatively, if you need to isolate some part of the waste stream from the rest of it—for example, if somebody in the house is ill and you don't want to cycle their wastes into your garden soil, or if you want to collect and concentrate urine as a rich source of fertilizer—you can construct a separate cycle that takes the separate waste stream in a different direction and subject it to different transformations, so that whatever cycles back around to you is a resource rather than a problem.

Notice, however, that one of the limitations on energy discussed in the previous lesson also applies to matter: matter doesn't move in circles all by itself. It needs energy, information, and additional matter to ride that cycle back around to you. In one case, the energy that moves the matter in its circle is the gravity that makes water flow downhill; the additional matter is the water of the river where your town's sewage plants discharge their waste, and the information is the biological programming that makes the fish you're going to eat next week recognize and snap up the aquatic critters that just ate little bits of your feces. In another case, the energy is the heat released by thermophilic bacteria in your compost pile; the additional matter is the garden and kitchen waste that makes up most of the pile, and the information is partly tucked away in bacterial DNA and partly your own knowledge of how to run a compost bin. Either way, the same limitation applies. Much more often than not, it's these additional requirements, rather than the nature or available quantity of the matter you're working with, that imposes hard limits on what you can accomplish.

The same principles that allow you to make the cycle run the way you prefer, and the same requirements of energy, matter, and information, apply equally wherever matter is involved in the green wizard's work. It's crucial to recognize, though, that not everything can be transformed into a useful form, or even a harmless one, by guiding the circles in which it moves. One of the essential boundaries of appropriate tech is the boundary between the kinds of matter you can change with the tools you have on hand, and the kinds you can't, and if you can't change it into something safe, it's a bad idea to produce it in the first place. It really is that simple: If you can't transform it, don't produce it.

Exercise for Lesson 2

Take at least one material item or substance you currently get rid of, and figure out where it goes once it leaves your possession. Don't cheat yourself by choosing something you already know about, and don't settle for abstractions. With the internet at your fingertips, it takes only a modest amount of work to find out which landfill gets your garbage, which river has to cope with your sewage, and so on. Your ultimate goal is to trace your chosen item or substance all the way back around to your own front door—for example, by tracing your plastic bottles to a particular landfill, the polymerizers in the bottles to the groundwater in a particular valley, the groundwater to a particular river, and the river to the particular coastal waters where the local fishing fleet caught the fresh cod you're about to have for dinner.

This may be an unsettling experience. I apologize for that, but it can't be helped. One of the few effective immunizations against the sort of airy optimism that assumes that pollution can't hurt us is to spend time wrestling with the muddy, material details of our collective predicament. If your wizardry is going to amount to more than incantations that make people feel better about themselves even as society consumes its own future, it needs to get into the nitty gritty of the work—first with the mind, then with the hands.

✒︎ LESSON 3 ✑
Introducing Information

Information is the third element of the triad of fundamental principles that flow through whole systems of every kind. All three need to be understood to build viable systems using appropriate technology. In making sense of this third factor, the appropriate tech movement had a huge advantage that people only a few decades earlier didn't have: the science of cybernetics, which is the science of information flow. It is one of the great intellectual adventures of the twentieth century, and it deserves much more attention than most people give it these days.

Unfortunately, in trying to pick up where the appropriate tech movement left off, we now have a huge disadvantage that people a few decades didn't have. The practical achievements of cybernetics, especially but not only in the field of computer science, have accidentally midwifed a set of attitudes toward information in popular culture that impose bizarre distortions on the way most people nowadays approach the subject. You can see these attitudes in their extreme form in the theory of the Singularity, promoted by computer scientist Ray Kurzweil and popular in some circles. That theory suggests that since the amount of information humanity has is supposedly increasing at an exponential rate, and exponential curves approach infinity asymptotically in a finite time, then at some point not too far in the future, industrial humanity will know everything and achieve something like godhood.[5]

Believers in this prediction insist that it's an avant-garde scientific concept backed by all sorts of data. In point of fact, it's nothing of the kind; it's simply a rehash of Christian apocalyptic myth in the language of cheap science fiction, complete with a techno-Rapture into a heaven lightly redecorated to make it look like outer space.

The entire notion of the Singularity is rooted in a basic misunderstanding of what information is and what it does. This is all the more embarrassing in that this misunderstanding was dealt with conclusively decades ago by the pioneering theorists of cybernetics. Gregory Bateson was one of those pioneers, and his work—which you would have found on the shelf of any serious appropriate tech geek back in the day—is a good place to start clearing up the mess. Bateson defines information as "a difference that makes a difference."[6] This is a subtle definition, much more subtle than it seems at first glance, and it implies much more than it states. Notice in particular that whether something "makes a difference" is not an objective quality possessed by a difference; it depends on an observer, to whom the difference makes a difference. To make the same point in the language of philosophy, information can't be separated from intentionality.

What is intentionality? The easiest way to understand this concept is to turn toward the nearest window. Notice that you can look *through* the window and see what's beyond it, or you can look *at* the window and see the window itself. If you want to know what's happening in the street outside, you look through the window; if you want to know how dirty the window glass is, you look at the window. The window presents you with the same collection of photons in either case; what turns that collection into information of one kind or another—what makes the difference between seeing the street and seeing the glass—is your intentionality.

The torrent of raw difference that deluges every human being during every waking second, in other words, is not information. That torrent is data—a Latin word that means "that which is given." Only when we approach data with intentionality, looking for differences that make a difference to us at that moment, does data become information—also a Latin word, meaning "that which puts form into something." Data that isn't relevant to a given intentionality, such as the dirt on a window when you're trying to see what's outside, has a different name, one that doesn't come from Latin: noise.

Thus the mass production of data in which true believers in the Singularity place their hopes of salvation is much more likely to have the opposite of the effect they claim for it. Information only comes into being when data is approached from within a given intentionality, so it's nonsense to speak of it as increasing exponentially in any objective sense. Data can increase exponentially, to be sure, but this simply increases the amount of noise that has to be filtered before information can be made from it. This is particularly true in that a very large portion of the data that's exponentially increasing these days consists of such important material as, say, internet gossip about the current color of Lady Gaga's pubic hair.

Data, in other words, must be filtered in order to turn it into information. The process of filtering data to produce information and get rid of noise, in turn, always requires energy, matter, and additional information. Like energy and matter, information doesn't function in a vacuum; whether the information is in a plant's DNA, a human brain, or an electronic device, it has to be stored in matter, by the application of some kind of energy, using some additional set of information that governs how it is encoded. As with energy and matter, in turn, it's usually these additional factors that impose hard limits on the kinds and amounts of information that can be used in any given situation.

Once information is created and used, in turn, it has to be dealt with. A buildup of too much information in a system is just as damaging as a buildup of too much energy or too much matter. A blackboard soon becomes useless if there's no eraser to clear away one set of markings and make room for another, and the equivalent rule applies to every kind of system: information has to be sorted for long-term relevance, and anything that is only relevant in the short term needs to be purged ruthlessly once its usefulness has expired. This is why so much of what goes into your short-term memory, for example, never makes it into long-term memory. Once again, energy, matter, and more information have to go into the purging process we call forgetting, and these requirements also impose limits on what can be done with information.

The need to filter data so that information can be extracted from it explains why the sense organs and nervous systems of living things have been shaped by evolution not to expand but to restrict the kinds and volumes of data they accept. Every species of animal has different information needs, and each limits its intake and processing of data in a different way. You're descended from mammals that spent a long time living in trees, for example, which is why your visual system is very good at depth perception and seeing colors (important in differentiating ripe from unripe fruit) and very poor at many other things.

A honeybee has different needs for information, so its senses select different data. It sees colors that extend well up into the ultraviolet, which you can't, because many flowers use reflectivity in the ultraviolet to signal where their nectar is. The bee also sees the polarization angle of light, which you don't, since this helps it navigate to and from the hive. You don't "see" heat with a special organ on your face, the way a rattlesnake does, or sense electrical currents the way many fish do. Around you at every moment is a world of data that you will never perceive. Why? Because your ancestors, over millions of generations, survived better by excluding that data, so they could extract information from the remainder, than they would have done by including it.

Human social evolution parallels biological evolution, so it's not surprising that much of the data processing systems in human societies are ways of excluding most of the available data so that useful information can emerge from the little that's left over. This is necessary, but it's also problematic: a set of filters that limit data to what's useful in one historical or ecological context can screen out exactly the data that might be most useful in a different context, and the filters don't necessarily change as fast as the context. One very common consequence of the lag time between social change and human cultural filters is the emergence of persistent logical mistakes that cripple a society's ability to deal with changing times.

Belief in the Singularity, and less obviously religious forms of faith in perpetual progress, are based on exactly such a logical mistake. Believers in these notions have forgotten that the words "it can't be done" also convey information, and a very important kind of information at that. It's entirely plausible that even if we did achieve infinite knowledge about the nature of the universe, what we would learn from it is that the science fiction fantasies being retailed by believers in the Singularity are out of reach, and we simply have to grit our teeth and accept the realities of human existence after all.

Information always implies an intentionality because data doesn't turn into information until it passes through a filter oriented toward a particular intentionality. Even something as simple as a thermostat connected to a furnace has an intentionality—the thermostat "looks" at the air temperature, and "sees" if that temperature is low enough to justify turning the furnace on, or high enough to justify turning it off. The better the thermostat, the more completely it ignores data that has no bearing on its intentionality. Conversely, most of the faults thermostats can suffer are ways that other bits of data slip through the filter (for example, the insulating value of the layer of dust on the thermostat) and insert themselves where they're not wanted.

The function of the thermostat in the larger context to which it belongs—the system of the house that it's supposed to keep at a more or less stable temperature—is another matter, and it requires a subtly different intentionality. The homeowner, whose job it is to make information out of the available data, monitors the behavior of the thermostat and, if something goes wrong, has to figure out where the trouble is and fix it. The thermostat's intentionality is to respond to changes in air temperature; the homeowner's intentionality is to make sure that this intentionality produces the effect that it's supposed to produce.

One way or another, this same two-level process plays a role in every part of the green wizard's work. It's possible to put additional levels between the information filter on the spot (in the example, the thermostat) and the human being who manages the system; but in appropriate tech it's rarely a good option. The Jetsons fantasy of the house that runs itself is one of the things that most deserves a place in history's compost heap as the age of cheap energy comes to a close. As a green wizard, your goal in crafting systems is to come up with stable, reliable information filters that will pursue their own intentionalities without your interference most of the time, while you monitor the overall output of the system and keep tabs on the very modest range of data that will let you know if something has gone haywire.

Notice what's happening in the two levels of processing in the example of the thermostat: the total amount of data is being filtered in two ways, one narrow and one broad, so that useful information can be separated from noise. The thermostat focuses precisely on the narrowest set of data that will allow it to do its job, while the homeowner pays attention to the overall comfort level in the house, as well as to such details as how often the furnace cycles on and off, and how much the monthly fuel bill is running. Machines are good at tight filtering of data to produce precise but narrow flows of information; human beings are good at broad overviews of data to produce imprecise but global flows of information. Both these flows are needed to keep a system working—and the privileging of narrow, machine-filtered flows of information over broader, human-filtered flows goes a long way to explain why industrial civilization is facing the rising spiral of crises that dominates today's headlines.

> ### Exercise for Lesson 3
>
> Take at least one flow of information that shapes the way you live your life, and trace the following factors: the underlying data that has to be filtered in order to extract information, the thing that does the filtering, the intentionality of the filter, and the means for disposing of old information no longer relevant to the intentionality. This will be much easier for you, especially at first, if you choose something as narrowly specific as the thermostat example: for instance, a set of data you have to monitor at your job, or the specific variations in local weather that affect how you spend your weekends.
>
> Once you've identified the flow and explored the factors just listed, see if you can find the two kinds of information filtering described above: the narrow and precise kind that is often best handled by a machine (such as the temperature and rainfall measurements during the week) and the broader and imprecise kind that is usually best handled by a human being (such as the forecast of the weekend's weather). If one of these is lacking, think of ways in which it could be introduced into the way that the information is handled.

LESSON 4
Thinking in Systems

The three factors just outlined—energy, matter, and information—are basic to an ecological understanding of the world, and they form the threefold foundation of the green wizardry of appropriate tech. A foundation isn't a building, though, and it's entirely possible to start from a good working knowledge of one or more of the three factors and still end up with something that's hopelessly unworkable in practice. The blueprint that makes it possible to construct a viable building on that foundation—to extend the architectural metaphor a little further—is a solid, basic knowledge of what systems are and how they work.

What is a system? In the 1960s and 1970s, when systems theory was in its golden age, students of appropriate tech had their choice of any number of carefully drawn definitions. For practical purposes, though, we can condense those definitions down to the following: a system is any set of interacting components in which the behavior of the whole set is determined by the sum of the interactions among its components.

Got that? Now let's go through it in detail.

A system is a set—that is, it consists of two or more things, which we'll call components of the system. Those components are interacting—that is, each of them acts on other components of the system, and each of them responds to the actions of other components. The whole system has its own behavior, which is distinct from the individual actions and responses of the components of the system. To use some systems jargon, the behavior is an *emergent property* of the system; this means that it emerges out of the sum of all the actions and responses of the components to each other, rather than out of any individual component.

The thermostat we examined in the previous lesson sits at the core of a very simple system. The components of the system are the thermostat, the furnace, and the air inside the house. The thermostat measures the temperature of the air, and turns the furnace on and off. The furnace receives signals from the thermostat and either heats or doesn't heat the air. The air receives heat from the furnace, and the heated air affects the behavior of the thermostat. None of the individual components of the system—the thermostat, the furnace, or the air—is responsible for the behavior of the whole system; all three of them must work together to produce the effect of the whole system, which is to keep the air temperature inside the house close to some desired value.

This very simple system has two features that are common to every other viable system. The first is that the thermostat system is not and cannot be isolated from its surroundings; it can only function because it receives inputs from, and returns outputs to, its surroundings. The thermostat can signal the furnace only because electricity carries the signal; the furnace can heat the air only because it receives fuel and vents combustion wastes up the chimney; the air is able to hold an even temperature only because the walls of the house keep the heat from diffusing itself out into the atmosphere as a whole. Thus the system of thermostat, furnace, and air is a subsystem of a larger system, and that larger system is a subsystem of an even larger system. Trace the inputs and outputs of any system far enough, and they extend to the limits of the universe.

It's sometimes convenient to think of systems as having neatly defined edges, but this habit of thought has its downsides. What systems have, more precisely, are interfaces, which function very much like the walls of a cell or the borders of a country: that is, they limit the interactions between what's inside and what's outside in ways that tend to preserve the system on the inside. There are normally several different interfaces surrounding any system, each one limiting a different set of interactions between the system and its surroundings. It can be useful to think of interfaces as hard limits, in order to concentrate on the behavior of whatever system you're trying to understand, but that narrow focus can be a potent source of disasters if what's outside gets ignored too often. The frequency with which this happens can be credited to the simple fact that most people in the world's industrial nations are stupid about systems.

There's really no gentler way to put it. Listen to the media, the internet, or everyday talk these days, and you're guaranteed to hear somebody insisting that worrying about the limits to growth is just plain silly because science, technology, progress, the free market, the space brothers, or some other convenient *deus ex machina* will let us keep extracting an ever-increasing supply of energy and raw materials from a finite planet without ever running short, and always find places to dump the rising tide of waste without having it turn up again to give us problems. Now, of course, much of that rhetoric comes from flacks-for-hire pushing the agenda of various corporate or political pressure groups, but the illogic is pervasive enough that I suspect a lot of it comes from ordinary people who basically haven't yet noticed that the world isn't flat.

Watch what passes for political and economic debate these days, and you can count on hearing the same thing. Take "sustainable growth," the mantra of the business-as-usual end of the green movement. Even the most elementary grasp of systems theory makes it clear that there's no meaningful sense of the adjective "sustainable" that can cohabit with any meaningful sense of the noun "growth." In a system—any system, anywhere—limitless growth is always unsustainable; every system that works, anywhere in the cosmos, has mechanisms to prevent limitless growth. This is the second feature the very simple system just described shares with all other viable systems: it embodies negative feedback.

The phrases "negative feedback" and "positive feedback" are among the most misunderstood terms in systems theory, which is saying something. Negative feedback is not negative in a moral sense—it's not "bad feedback"; it's negative in the sense that it says "no" to changes in an environment. When the temperature of the air goes up too far, the thermostat turns off the furnace and lets it cool down; when it cools down too far, the thermostat turns the furnace back on and heats it up. Positive feedback works the other way: a thermostat set to positive feedback would respond to rising temperatures by turning on the furnace and heating the air even further. The result would be a house that's unbearably hot, and, if nothing keeps the temperature from rising indefinitely, the house will eventually catch fire and burn to the ground. Where feedback is concerned, in other words, positive isn't good and negative isn't bad; systems that have negative feedback sustain themselves, while systems with positive feedback destroy themselves.

The modern industrial economy is a system with positive feedback—that's what "limitless economic growth" means in systems terms—and therefore, with mathematical certainty, it will destroy itself unless negative feedback kicks in first. Climate instability is part of that negative feedback because an industrial society ravaged by droughts, floods, tornadoes and hurricanes will experience less growth than it would in the absence of these little problems. The depletion of fossil fuel reserves is another form of negative feedback; an industrial society that has to put a rising portion of its available resources into extracting fossil fuels from the ground will experience less growth than it would if it somehow got a limitless supply of energy from somewhere. How hard will these and other forms of negative feedback hit the industrial world? Over the next few decades, we're going to find out.

A basic grasp of systems thinking would make it easier to avoid such follies, but that same grasp would also make it impossible to pretend that we can go on living our current lifestyles much longer. That's an important reason why systems thinking was dropped like a hot rock in the early 1980s and why, outside of a narrow range of practical applications where it remains essential, it's been shut out of the collective conversation about our society ever since. For the aspiring green wizard, on the other hand, there are few habits of thought more important than thinking in whole systems. Most of what this book covers, even those things that seem strictly practical in nature, are explorations in systems thinking. Most of the work of green wizardry is simply the application of a systems approach to some aspect of working with nature.

The points already made should be applied to any system you are trying to understand. Start with the main components of the system and trace out the way they interact with one another; figure out where the interfaces are that manage the system's interactions with the rest of the universe; track the negative feedback that keeps the system from straying outside the conditions that allow it to function. Then work out how energy flows through that system, how matter cycles into and out of that system, and how data is filtered by that system to separate the information the system needs from the noise it needs to exclude. Once you know all these, you'll be in a good position to make sense of the system. If this reminds you of the exercise at the end of Lesson 3, it should; the habit of sorting out systems behavior suggested by that exercise is worth cultivating throughout your work as a green wizard.

In the case of the thermostat system we've been discussing, all these factors can be sketched out easily enough. We've already discussed its boundary zones and the way it relies on negative feedback. Energy comes into the system directly in the form of electricity and whatever fuel the furnace uses, and energy leaves the system by way of diffuse heat leaking out from the house into the outside air and heat going up the chimney. Matter comes into the system partly via the manufacturing, distribution, and installation processes that put the system in place, and partly as the fuel the furnace is burning—fuels are substances that contain energy, remember, and have to be included in both categories. Matter leaves the system partly when the system components are scrapped at the end of their useful lives, and partly in the form of carbon dioxide and other pollutants released by burning the fuel. The filtering of data into information follows the twofold pattern already noted.

Other systems are much more complex, and it may take you quite a bit of careful attention to figure out, for example, the tangled flows by which energy, matter, and information structure a backyard garden plot. Still, understanding the basic patterns will help keep you from making a number of very common mistakes, and they will help each of your projects in green wizardry achieve the negative feedback that will keep it in balance with its surroundings.

(**Exercise for Lesson 4**)

For this lesson's exercise, your task is to take a single system and trace the flows of energy, matter, and information that keep it functioning. The system in question is you.

You are a system, after all, in every sense of the word. You are a set of interacting subsystems—organs, tissues, and so on—and the interactions among these subsystems give rise to behavior that none of the individual subsystems can produce on its own. Negative feedback of various kinds keeps you alive and functioning, and boundary zones mediate your interactions with the world around you. Energy enters you in the form of food, and leaves you in the form of waste heat; matter has its own routes through you; and you filter data constantly to extract the information you need from a great deal of background noise.

Trace as many of these systems processes in yourself as you can by drawing a diagram, writing out a description, or in any other way that works for you. Make this as complete as you can manage, using whatever reference sources you need.

LESSON 5
Flows and Funds

The points covered in the first four lessons are meant to give you a very basic grasp of the ways that whole systems function in the real world. If you're interested in learning more—and you should be, if you want to learn appropriate tech—the books on ecology and systems thinking in the Resources section at the end of Part One will give you a good start on the theory, and working your way through the rest of the lessons in this book will give you an equally good start on the practice. Still, there are two more points that ought to be covered before we finish talking about principles, because they bear directly on the challenges that a revived appropriate technology movement will need to face as the age of cheap energy comes to an end.

The first of these points is the difference between flows and funds. It's a simple, straightforward distinction that most people in today's industrial societies use constantly in their daily lives. Try to suggest that they ought to apply the same distinction to energy, raw materials, or information, though, and you can count on getting either a blank look or a diatribe, because next to nobody these days wants to think about the consequences of applying common sense to resource issues.

The relation between flows and funds is complex, but it can be simplified in a useful way by using money as a metaphor. You can think of a flow as a paycheck. If you make five hundred dollars a week in take-home pay, that's how much you have available to spend from one Friday to the next, and if the potential uses for that money add up to more than five hundred a week, you have to prioritize. So much has to be set aside for rent, so much for food, so much for utilities, and so on, before you decide how much you can spend when you go out on Saturday night. Neglect that limit, and you can end up scrambling to get by until Friday comes around and you get your next paycheck.

You can think of a fund, in turn, as a lump sum of money, such as a lottery prize. If your lottery ticket wins you the jackpot, and suddenly you have ten million dollars in the bank, you may think that you don't have to prioritize at all. Still, unless you have money coming in from some other source, that ten million dollars is all you have to spend, and unless there's less than ten million dollars worth of expenditures you want to make over the rest of your life, you have to prioritize, just like the guy making five hundred a week.

Notice, though, that if you have a fund rather than a flow, there's a strong temptation to ignore priorities and run amok with your wealth, because you don't have to worry about the limits to a fund until the fund's nearly gone. The limit to a flow is a constant factor;

each week, you've got just five hundred dollars to spend, but seven days later you'll have five hundred more. The limit to a fund is not a constant factor, and it only affects you late in the game—but when it does show up, unless you've been unusually prudent, it clobbers you.

There are at least two ways to dodge the downside of using up a fund, and both involve turning it into a flow. The first is to calculate how long you're going to need to live off the fund, figure out how much you can take from the fund each week and still have some left as long as you need it, and then stick to that budget. The second method is to invest at least some of the fund in assets that give you a steady return, make the resulting flow the bedrock on which you build your financial life, and then don't touch the principal. Both those choices require you to accept hard limits now, in order to avoid bankruptcy when the fund runs out.

You can also convert a flow into a fund. Imagine that you have a small flow—say, an investment that pays a modest sum every month. If you put that money into a savings account and let it accumulate, you can find yourself with a nice lump sum after a few years—which can pay for things you might not be able to afford with your weekly paycheck. As the flow turns into a fund, though, the same temptations that affect any other fund apply to it, and it may take a certain amount of self-control to avoid running amok with it.

Apply the model of funds and flows to human ecology, and you've got the history of the modern world. Until our species broke into the Earth's store of fossil fuels and started going through it like a lottery winner on a spree, we lived from paycheck to paycheck on the energy flow from the sun and the material and information we could extract and maintain with that much energy. We got fairly good at it, too. Growing crops, raising livestock, building windmills and waterwheels, designing houses to soak up solar heat in winter and shed it in the summer, and many other ingenious tricks gave us the annual paycheck we used to support ourselves and cover the costs of such luxury goods as art, literature, philosophy, science, and the occasional Great Pyramid.

During that time, nature turned a variety of flows into funds—for example, converting the steady flow of sunlight through photosynthesis into forests and topsoil. All too often, human societies stumbled across these funds and ran amok with them, wasting them casually, disrupting the natural systems that produced them, and then running into deep trouble when the funds were exhausted. Some human societies learned the hard lesson taught by these experiences and figured out how to draw wealth out of nature's funds without exhausting them; other human societies failed to do so, and went extinct.

With the transformation of coal from ugly black rock to energy resource over the course of the eighteenth century, the human relationship to ecological flows and funds changed radically. Our species won the lottery, and it wasn't a paltry ten million dollar prize, either—it was the great-great-grandmother of all jackpots, half a billion years of stored sunlight in highly concentrated and easily accessible forms. For most of three hundred years, the main constraint on how fast we burnt through the Earth's fossil fuel supply was how fast we could find clever ways to use it. What nobody noticed at the time—or for a long time thereafter—was that we'd switched from a flow to a fund, and the faster our fossil fuel use accelerated, the faster the bank balance depleted.

We could have done the smart thing and converted the fund into a source of flows. That was a good part of what the appropriate tech movement of the 1970s was trying to do: figuring out ways to use the world's remaining fossil fuel reserves to bridge the gap to a renewable energy technology that could last after the fossil fuels were gone. Even then, it was a gamble; nobody knew for sure if it would be possible, with the world's remaining fossil fuel reserves, to create a renewable infrastructure sturdy and productive enough that it could keep providing ample energy into the far future. Still, it was probably the best chance that we had left.

A great many people these days insist that a massive buildout of green technology is still an option. It isn't, and the reason for that comes from a core difference between money, on the one hand, and the three basic factors of ecology—energy, matter, and information—on the other. You can take money out of a bank account at whatever rate you wish, and for all practical purposes there's no cost to doing so, except for that final accounting when you can no longer cover your bills. It doesn't work that way with energy, matter, or information: the more of these that you want, the more energy, matter, and information you have to invest in the process. These costs can't be evaded, and they have to be factored into any decision in the real world.

There's a boobytrap here, though, because these costs behave in one way when applied to flows and in another, much more dangerous way, when applied to funds. For example, the annual supply of sunlight to the Earth is a flow. (Strictly speaking, that flow comes from a fund—the fund of the sun's reserves of hydrogen—but in any timescale meaningful to human beings, it's a flow.) The amount of energy, matter, and information you need to invest in order to tap into that flow by building a solar water heater, say, is roughly the same now as it was in the 1970s, and it'll be roughly the same twenty million years from now, too, when intelligent beings descended from raccoons get to work building their own solar water heaters.

As a general rule, in other words, the amount of energy, matter, and information you need to invest to tap into a flow remains more or less stable over time. The amount of these things you need to invest to tap into a fund, by contrast, does *not* remain stable over time; the cost increases as the fund depletes. For example, rock phosphate, a vital ingredient in chemical fertilizers, is one of the essential resources for industrial agriculture. Over any timescale relevant to human beings, the total amount of rock phosphate that can be mined is a fund, not a flow, and the geological processes that lay down deposits of rock phosphate follow what's called the *power law*—the rule, applicable across an astonishing range of natural phenomena, that whatever's ten times as large, or ten times as concentrated, is by and large ten times as rare.

What this means in practice is that rock phosphate, like every other natural resource, occurs in a small number of very large, very concentrated deposits, a moderate number of midsized and modestly concentrated deposits, and a very large number of very small and poorly concentrated deposits. All other things being equal, the big, concentrated deposits get discovered and extracted first, since they are both easier to find and more lucrative to extract; once those are exhausted, the industry moves to the next largest and most concentrated deposits, and so on down the scale until little pockets of rock with a bit of phosphate in them are all that's left.

Thus, the cost in energy, matter, and information needed to extract a pound of rock phosphate goes up over time, because it takes more of each of these things to find, dig up, and process a pound of phosphate from those little pockets than it did to get a pound of rock phosphate from the huge and highly concentrated deposits that were available at the beginning. The same thing is true of energy resources that exist in the form of funds, and it's equally true of funds of information—for example, scientific research is subject to a close equivalent to the power law; there are always some facts about nature that are easy to find, others that are harder, and still others that are theoretically learnable but that no one in the history of the universe will ever manage to learn because it would take the gross domestic product of a good-sized galaxy to fund the research program needed to discover them.

This is the point that's missed by all those loudly ballyhooed claims that the industrial world can simply keep on extracting raw materials from more and more diffuse sources, until finally we're feeding ordinary rock and sea water into our factories and pulling out individual atoms of whatever we want. In a purely abstract sense, that's possible; in the real world, long before you get to that point, the world's annual production of energy, economically useful matter, and information would have to be put into the job of extracting raw materials, with nothing left over to do anything with the raw materials once you got them. In terms of our money metaphor, that's when you go bankrupt; in terms of the real world, long before that point arrives, the skyrocketing costs of raw materials force the global economy to its knees.

If you're going to transform a fund into a flow, as a result, you have to do it before you reach the point where the costs of extraction from the fund swallow up all the leeway you have. This is why a grand transition to a green industrial future isn't going to happen at this point: the costs of extracting fossil fuels and most other energy resources—costs, that is, in terms of energy, matter, and information, not in terms of the easily manipulated tokens we call money—are climbing steeply, and so are the costs for most of the raw materials essential to industrial society, from rock phosphate to the rare earth minerals that go into high-tech gadgetry. There isn't enough leeway left to allow for the transition without bankrupting the global economy in the process.

This is one of the reasons why the lessons of the appropriate tech movement are so relevant now. Many of the pioneers of that movement deliberately turned away from the grand fantasy of a vast top-down transition to green industrialism, and instead embraced change on a scale that can still work today: the local, personal, inexpensive, do-it-yourself end of building a green society. That approach remains an option because growing at least some of your own food, making the most of modest amounts of energy, and the other practices in the appropriate tech toolkit cost less energy, matter, and information than the ways that people feed themselves and use energy today. The logic is the same as in the metaphor of money: if you depend on a fund for survival, and the fund is running out, cutting your expenses until you can meet them with whatever flows are available to you—no matter how modest those flows may be—is the one and only way to avoid disaster.

(**Exercise for Lesson 5**)

It's crucial to understand funds and flows on the large scale, but the same principle plays just as important a role on the scale of a backyard garden or an insulated home. Your assignment for this lesson, therefore, is to choose three forms of energy, three forms of matter, and three forms of information that support the way you live today, and figure out whether each of them comes from a fund or a flow. For each fund that supports your lifestyle, think of a flow that you could use to replace it—not in some abstract future when somebody else will hand you the necessary technology, but here and now, with the tools and technologies you yourself can afford.

⋘— LESSON 6 —⋙
Sustainability and Resilience

The principles already covered provide most of the grounding you'll need to venture into the world of green wizardry, but one more point needs to be made before we put on overalls, boots, and work gloves and head for the backyard to get to work. That point, oddly enough, depends on the awkward process of remembering what a couple of today's popular buzzwords actually mean. This process is more than a little reminiscent of fishing scrap metal out of a swamp; in the present case, the words that need to be hauled from the muck, hosed off, and restored to their former usefulness are "sustainability" and "resilience."

Both of these terms have seen heavy use as rhetorical weapons in recent years, and they've come through the experience much the worse for wear. Sustainability was picked up first, but turned out to have some highly problematic dimensions. To be sustainable, something—a technology, a lifestyle, or what have you—has to be able to keep going indefinitely no matter what the future throws at it. Two things the future can throw at it deserve particular attention in this context. The first is *ecosystem limits*. If a technology that's supposedly sustainable depends on using nonrenewable resources, for example, or on using otherwise renewable resources at a rate that exceeds the biosphere's ability to renew them, it's just flunked its sustainability test. In the same way, if a technology puts things into the biosphere that disrupt the natural cycles of matter, energy, and information that keep the biosphere going, it's not sustainable no matter how much green spraypaint you apply to it.

Less often grasped, because of its unwelcome implications, is something else the future can throw at a technology, which takes the form of *complexity limits*. This category sums up the relation between a supposedly sustainable technology and the social, economic, and technical dimensions of human society, now and in the future. If those systems have a significant chance of dropping below the level of complexity that your supposedly sustainable item needs in order to stay fueled, maintained, stocked with spare parts, and the like, no matter how green it looks or how enduring it might be in the abstract, it's not sustainable. Since most of today's supposedly sustainable technology can't be manufactured or repaired without a fully functioning industrial society powered by cheap abundant energy, complexity limits are a massive issue, and this awkward fact started to sink in about the time "sustainability" peaked as a buzzword.

Around the beginning of 2012, therefore, the word "sustainability" began to lose its privileged place in the jargon of the time, as it began to sink in that no matter how much manhandling was applied to that much-abused term, it couldn't be combined with the phrase "modern middle-class lifestyle" without resulting in total absurdity. Enter "resilience," which was enthusiastically embraced as another way to talk about what too many people nowadays want to talk about (to the exclusion of more useful conversations): the pretense that a set of lifestyles, social habits, and technologies that were born in an age of unparalleled extravagance can be maintained as the material basis for that extravagance trickles away.

The word "resilience" is guaranteed to suffer the same fate as "sustainability," and for exactly the same reason: it has a perfectly clear meaning. So far, nobody's been rude enough to talk about that meaning, so the green blogosphere is awash with talk about resilient this and resilient that. Once people start to figure out what resilience actually means, though, it's a safe bet that they'll be hunting for another buzzword in short order, because resilience can be defined very precisely: it's the opposite of efficiency.

Now that you've stopped spluttering, let me explain.

We can define efficiency informally but usefully as the practice of doing the most with the least. An efficient use of a resource—any resource, whether we're discussing energy, matter, or information—is thus a use that puts as little of the resource in question as possible into places where it sits around doing nothing. The just-in-time ordering process that's now standard in the retail and manufacturing sectors of the economy, for example, was hailed as a huge increase in efficiency when it was introduced; instead of having stockpiles sitting around in warehouses, items could be ordered electronically so that they would be made and shipped just in time to go onto the assembly line or the store shelf. What few people asked is what happens when something goes wrong.

This is not a minor problem, as every significant natural disaster in an industrial region reminds us. In the inefficient old days, when parts jobbers scattered all over the industrial world had warehouses full of parts being produced by an equally dispersed array of small factories, such events were only a problem for those directly affected; stocks of spares could satisfy demand until sidelined factories recovered. However, now that production is efficiently centralized in very few factories, or in some cases only one, and warehouses full of parts have been rendered obsolete by efficient new ordering systems, delays and resulting costs that would have been negligible in 1970 are proving to be very substantial today.

Efficiency, in other words, is not resilient. What makes a system resilient is the presence of unused resources, and these are inefficient, by definition. A bridge is resilient, for example, if it contains more steel and concrete than is actually needed to support its normal maximum load; that way, when some outside factor such as a hurricane puts unexpected stresses on the bridge, the previously unnecessary structural strength of the extra steel and concrete comes into play and keeps the bridge from falling down. Most bridges are designed and built with that sort of inefficiency in place because the downside of too little efficiency (the bridge costs more to build) is less troubling than the downside of too little resilience (the bridge collapses in a storm). Like every project worth doing, a good bridge has to strike a balance between many conflicting factors, no one of which can be maximized except at the expense of others of equal importance.

That may seem straightforward, but there's another dimension here that nobody likes to talk about. There's a reason why contemporary industrial culture is obsessed with efficiency, and it's not because we're smarter than our grandparents. Every civilization, as it nears the limits of its resource base, has to deal with the mismatch between habits evolved during times of abundance and the onset of shortages driven by too much exploitation of that abundance. Nearly always, the outcome is a shift toward greater efficiency. Localized governments give way to centralized ones; economies move as far toward mass production as the underlying technology will permit; precise management becomes the order of the day; waste gets cut and so, inevitably, do corners. All this leads to increased efficiency and thus decreased resilience, and sets things up for the statistically inevitable accident that will push things past the limits of the civilization's remaining resilience and launch the downward spiral that ends with sheep grazing among ruins.

I've come to think that a great many of the recent improvements in efficiency in the industrial world derive from this process. Loudly ballyhooed as great leaps forward, they may actually be signs of the tightening noose of resource constraints that, in the long run, will choke the life out of our civilization. Although it's a great idea in the abstract to demand a society-wide push for resilience, in practice, that would involve loading a great many inefficiencies onto the economy. Most things would cost more, and fewer people would be able to afford them because the costs of resilience have to be paid and the short-term benefits of excess efficiency would have to be foregone. That's not a recipe for winning an election or outcompeting a foreign rival, and the fact that it might just get us through the waning years of the industrial age pays nobody's salary today. It may well turn out that burning through available resources, and then crashing into ruin, is simply the most efficient way for a civilization to end.

Where does that leave those of us who would like to find a way through the crisis of our time and hand down something of value to the future? This is a less challenging issue, since individuals, families, and local communities often have an easier time looking past the conventional wisdom of their era and doing something sensible even when it's not popular. The first thing that has to be grasped is that trying to maintain the comfortable lifestyles of the recent past is a fool's errand. It's only by making steep cuts in our personal demand for resources that it's possible to make room for inefficiency, and therefore resilience.

Most of the steps proposed in this book are inefficient. It's inefficient in terms of your time and resources to dig up your backyard and turn it into a garden, but that inefficiency means that if anything happens to the hypercomplex system that provides you with your food—a process that reaches beyond growers, shippers, and stores to the worlds of high finance, oil production, resource politics, and much more—you still get to eat. It's equally inefficient to generate your own electricity, to retrofit your home for conservation, to do all the other things that will be discussed in this book. Those inefficiencies, in turn, are measures of resilience; they define your fallback options, the extra strength you build into the bridge to your future, so that it can hope to stand up to the approaching tempests.

One of the crucial details of this approach involves the hugely unpopular step of relying on older technologies instead of their up-to-date equivalents. By and large, older technologies are less efficient. That means, in turn, that older technologies tend to be more resilient. Using those technologies means accepting downscaled expectations; a tube-based radio is easy, a tube-based television is challenging, and a tube-based video game would be about the size of a mobile home and use as much power as a five-story office building. This is why, sixty years ago, radios were cheap, televisions were pricey, and games were played on brightly colored boards on the kitchen table without any electronics at all.

Downshifting to that less complex approach may be a bit of a challenge for those raised to believe that technological progress only goes one way. Still, downscaled expectations will be among the most common themes of the decades ahead of us, and those who have the uncommon sense to figure this out in advance and start getting ready for a less efficient future will very likely benefit from their increased resilience.

Exercise for Lesson 6

For this lesson's exercise, go back to the three forms of energy, three forms of matter, and three forms of information you listed in the exercise for Lesson 5. Ask yourself what you would do if your access to each of them was suddenly interrupted (a) for one day, (b) for two weeks, or (c) forever. What would you have to have on hand to weather those interruptions without serious crisis? How inefficient would it be for you to have those things on hand?

RESOURCES FOR PART ONE

Our time, as the media never tires of telling us, is the information age, an era when each of us can count on being besieged and bombarded by more information in an average day than most premodern people encountered in their entire lives. While this is true in one sense, it's important to remember that it only applies if the meaning of the term "information" is restricted to the sort of information that comes prepackaged in symbolic form. The average hunter-gatherer moving through a tropical rain forest picks up more information about the world of nature through his or her senses in an average day, after all, than the average resident of an industrial city receives through those same channels in the course of their lives.

Still, the information the hunter-gatherer receives is the sort that our nervous systems, and those of our ancestors back down the winding corridors of deep time, evolved to handle. The current glut of symbolic information—words and images detached from their organic settings and used as convenient labels for abstractions—is quite another matter. There are certain advantages to the torrent of abstract information available to people in the industrial world these days, to be sure, but there's also a downside, and one major part of that is a habit of shallow thinking that governs most of our interactions with the information around us.

A hundred years ago, a student pursuing a scientific or engineering degree might need half a dozen textbooks for the entire course of his studies. Every chapter, and indeed every paragraph, in each of those books would be unpacked in lectures, explored in lab work, brought up in tests and term papers, so that by the time the student graduated, he had mastered everything those textbooks had to teach. That kind of study is almost unheard of nowadays, when students shoulder half a dozen huge textbooks a term, and have so little time to process any of them that the bleak routine of memorize, regurgitate, and forget is usually the only option. This is problematic in any sense you care to contemplate, but it has a special challenge for potential green wizards, because very few people these days actually know how to study information the way that a project of this nature demands.

Treat the material I'll be covering in this book as a collection of hoops to jump through on the way to some nonexistent degree in green wizardry, and the result will be abject failure. If you plan on studying this material, you need to pursue it with the same intensity the average twelve-year-old *Twilight* fan lavishes on sparkly vampires. You need to obsess about it the way a computer geek obsesses about obscure programming languages. You need to drench yourself in it until it shows up in your dreams and seeps into your bones.

You need to do these things because the ideas central to the old appropriate technology contradict the conventional wisdom of today's industrial cultures at every point. All the things that we learned about the world by osmosis, growing up in a society powered by cheap abundant fossil fuels and geared toward a future of perpetual progress, have to be unlearned in order to understand and use appropriate tech. It's precisely because many of us haven't yet unlearnt them that you see so many grand plans for dealing with peak oil that assume, usually without noticing the assumption, that all the products of cheap abundant energy will be readily and continuously available in a world without cheap abundant energy.

In my experience, the best way to begin this process is to go to a used book store. It needn't be the biggest and best in your area, if your area has more than one—in fact, you'll be more likely to get good results by going to one of those out-of-the-way places where the stock doesn't turn over too quickly and some of the books have been there for a good long time.

You're looking for books from the original appropriate technology movement of the 1970s and 1980s. There were thousands of books that came out of that movement, a small number from the large publishing houses of the time and a much vaster supply from small presses run by individuals, or by the little nonprofit groups that created so much of the appropriate tech movement in the first place. You may find anything from professionally bound hardbacks with dust jackets down to staplebound pamphlets with hand-sketched illustrations. You may find them in the gardening section, the home repair section, the science section, the nature section, or in a special section all its own labeled "Homesteading" or "Back to the Land" or something like that. (I haven't yet found a store that labeled it "Naked Hippie Stuff," but hope springs eternal.) You may even find it jumbled up all anyhow with the general nonfiction because the proprietor of the used book store had no idea where to put it.

Wherever it turns up, you're looking for books on organic gardening, conservation, renewable energy, or anything related to them. If you get one of the classics—*The Integral Urban House, Other Homes and Garbage, Rainbook, The Book of the New Alchemists, The Food and Heat Producing Solar Greenhouse*, or the like— that's good, but it's not required. It counts as much if you find a little staplebound pamphlet on composting, or a ragged trade paperback from a small press on building a solar oven, or an old Rodale Press hardback on insulated window coverings, or what have you.

There are two points to this exercise—well, actually, two and a half. The first point is that your work with green wizardry shouldn't be limited to what one middle-aged archdruid studied and practiced, under sometimes sharply limiting conditions, over the last thirty years or so. If you want to make green wizardry part of your life, a modest library of books on the subject is essential. You'll develop your own personal vision of appropriate tech, and your own personal style in putting it to work; the books you read and study, whether you agree with them or not, will help you start the process of bringing the vision and style into being.

The second point is that many of these books are nearing the end of their useful lives. The limited budgets available to most of the appropriate tech presses meant that most of the books were printed on cheap paper and bound by cheap methods. If they're going to become anything but landfill, and if the information they contain is going to find a new home, somebody needs to take responsibility for making that happen and, dear reader, it might as well be you.

The half point is that the appropriate tech movement had its own quirky culture and its own distinctive take on things. This is true of any specialized field. If you study a martial art, let's say, an important part of your early learning curve has to do with picking up the customs and traditions and little unspoken rituals of the art, which have nothing to do with how to block a punch and everything to do with navigating the learning process and interacting with your teachers and fellow students. This subtle learning remains important even when the movement no longer exists and the surviving participants have either gone on to other things or have spent the last thirty years laboring away in isolation at ideas and practices nobody else cares about; your task is harder, that's all, and one of the few ways you can get a sense of the culture of the movement is to spend time with its writings and its material products.

While you're at it, by the way, get a good basic book on ecology. Ecologist William Catton, whose book *Overshoot* was required reading back in the day, neatly defined ecology as the study of the processes that matter;[7] if anything, that's an understatement. If at all possible, the book you choose should be from the late 50s, 60s or 70s, when the concept of the ecosystem was made the central focus of study in a way that hasn't always been the case since then; the ecosystem approach is central to the way of thinking we'll be exploring in the pages to come, and a good grounding in it will be essential.

Each section of this book ends with a resource section listing some of the classic works from the old appropriate tech movement, or from the traditions and movements that helped launch it and guide it on its way. There have been plenty of good books on similar subjects written since the end of the movement, to be sure, but those latter are relatively easy to find. Many of the older books are not; a troubled collective conscience has shoved them down our culture's memory hole, and unless you happen to have an old copy of *The Whole Earth Catalog* or *Rainbook* handy, your chances even of finding out that they exist are not good. I have tried to include as many of the classics as possible, and a good selection of the merely useful books as well; still, plenty of gems will have been missed. What follows is simply my best attempt at what would have been considered, back in the day, a good collection of books on the basic principles of appropriate tech, with a few notes on each.

Bateson, Gregory, *Mind and Nature: A Necessary Unity* (New York: Bantam, 1979).

——, *Steps to an Ecology of Mind* (New York: Ballantine, 1972). One of the major figures in the emergence of cybernetics, Bateson had the gift of communicating complex systems concepts in readable language. *Mind and Nature* is his most important book; *Steps to an Ecology of Mind* collects many of his papers and essays.

Brewer, Richard, *Principles of Ecology* (Philadelphia: Saunders College Publishing, 1979).
A good introductory textbook of ecology.

Buchsbaum, Ralph and Mildred, *Basic Ecology* (Pacific Grove, CA: Boxwood Press, 1957).
Perfectly titled, *Basic Ecology* explains the core concepts of ecological thinking in terms that a bright eight-year-old can follow.

Catton, William R., Jr., *Overshoot: the Ecological Basis of Revolutionary Change* (Urbana, IL: University of Illinois Press, 1980). A constant intellectual companion in my college days, this is both a solid introduction to ecological thinking and a stark warning of the future into which industrial society has backed itself. If I could hold the entire American political establishment at gunpoint and make them read one book, this would be it.

Ehrlich, Paul R., Anne H. Ehrlich, and John P. Holdren, *Ecoscience: Population, Resources, Environment* (San Francisco: W. H. Freeman and Co., 1977).

A big textbook, *Ecoscience* covers most dimensions of the crisis of industrial society; while the details need updating, the basic concepts remain valid today.

Hardin, Garrett, *Filters Against Folly* (New York: Penguin, 1985).
Always controversial but always insightful, Hardin's writings have lost none of their value over the decades. This is the most useful of his books for the aspiring green wizard.

Johnson, Warren, *Muddling Toward Frugality* (San Francisco: Sierra Club Books, 1978).
Widely and deservedly praised, this book sketched out the strategy of local and personal change that undergirded most of the appropriate tech movement.

Kormondy, Edward J., *Concepts of Ecology* (Englewood Cliffs, NJ: Prentice-Hall, 1969).
Another good introductory textbook of ecology.

Laszlo, Ervin, *Introduction to Systems Philosophy: Toward a New Paradigm of Contemporary Thought* (New York: Gordon and Breach, 1972).
——, *The Systems View of the World* (New York: George Braziller, 1972).
Laszlo was probably the most intellectually ambitious of the 1970s systems theorists; the first of these books uses systems theory as the basis for a comprehensive philosophical stance, the second was among the standard textbooks in introductory systems theory classes.

Leopold, Aldo, *A Sand County Almanac* (New York: Oxford University Press, 1949).
If you were into appropriate tech back in the day, you inevitably read this classic of American ecological thought. If you want to learn about appropriate tech now, it's still an excellent place to start.

Meadows, Donnella, Dennis Meadows, Jorgen Randers, and William W. Behrens III, *The Limits to Growth* (New York: Universe, 1972).

Still the best guide to the future ahead of us, and thus inevitably the most thoroughly denounced of all the works in its genre. The several updates are also worth reading.

Odum, Eugene P., *Fundamentals of Ecology* (Philadelphia: W. B. Saunders, 1971).

The bible of ecosystem-based ecology, this used to be the standard textbook in upper-division ecology classes, and it beats the stuffing out of most modern books on the subject.

Roszak, Theodore, *Where the Wasteland Ends: Politics and Transcendence in Postindustrial Society* (Garden City, NY: Doubleday, 1972).

A brilliant critique of the mythology of progress, this was another of the books you could expect to find on most appropriate tech geeks' bookshelves.

Schumacher, E. F., *A Guide for the Perplexed* (New York: Harper Perennial, 1977).

——, *Small is Beautiful: Economics as if People Mattered* (New York: Harper, 1973).

Schumacher was the founder and patron saint of appropriate tech, and his books are as timely now as they were in the 1970s. *Small is Beautiful* provides the basic intellectual framework for appropriate tech; *A Guide for the Perplexed* is a deeper and more philosophical work, exploring the blind spots in modern thinking.

Weinberg, Gerald M., *An Introduction to General Systems Thinking* (Somerset, NJ: John Wiley & Sons, 1975).

If your introductory course on systems thinking back in the day didn't assign Laszlo's The *Systems View of the World* as its textbook, it used Weinberg's *Introduction*. A solid practical guide, worth detailed study.

PART TWO

Food

The Small Garden

Food and energy were the two core concerns of the appropriate tech movement in its heyday. Though the most innovative ventures of that movement combined the two, it's best to take them one at a time, so we'll begin with food. More precisely, we'll begin with the framework by which food is grown and distributed in a society that doesn't have fossil fuels to throw around casually. Fortunately, it's only been a hundred years or so since cheap energy twisted the industrial world's food system into its present unsustainable shape, so the basic pattern of the older and less extravagant system is still fairly easy to reconstruct.

If you hopped into a time machine and went back to visit farm country a century ago, in the days when sprawling interstate highway systems and fleets of trucks hadn't yet made distance an irrelevance over continental scales, you'd notice something about the farms of that time that you won't find in most farms today: each farm had, apart from its main acreage for corn or wheat or what have you, a kitchen garden, an orchard, a henhouse, and a bit of pasture for a cow or two. These had a completely different economic function from that of the main acreage, and they were managed in a completely different way. Their function was to produce food for the farm family and farmhands; they were worked *intensively* while the main acreage was farmed *extensively*.

The shift in prefixes between these two words points to two radically different ways of producing food. Extensive farming, as the term suggests, involves significant acreage. In the days before modern petroleum-based farming, at least, it maintained soil fertility through crop rotation and fallow periods, rather than through fertilizers or soil amendments. The basic tools of the trade are a plow and something to draw it—that meant horses or oxen, in the days before there were factories to produce tractors and fossil fuels to power them. The food plants you can grow with extensive farming in temperate regions, in the absence of cheap fossil fuel energy, are pretty much limited to grains and legumes, but you can produce these in huge amounts. They store and ship well, so they make good cash crops even if the only way to get them to market is a wagon to the nearest river system and a canal boat from there.

Intensive gardening has to be done on a much smaller scale; among other reasons, the labor it requires is too substantial to be applied to acreage of any size. It maintains soil fertility by adding whatever soil amendments are available—compost, manure, leaf mold, a fish buried in every corn hill, you name it—and the basic tools of the trade are a hoe and somebody who knows how to use it. The crops you can grow in an intensive garden account for everything other than grains and dry legumes, from the first spring radishes to the leeks you overwinter under straw. The chickens, the cow, and the fruit from the orchard also belong to the intensive sector and participate in its tight cycles of nutrients. In an age without fossil fuels, very little of what can be grown intensively can be transported over any distance without spoiling, so intensive growing is always done close to where the food will be eaten.

This is why every farm in the America of a century ago had

its own intensive kitchen garden, orchard, and livestock, and it's why every American city a hundred years ago was ringed with market gardens, egg farms, dairies, and the like, to keep the shelves of urban grocers filled with something besides grains and dried legumes. It's also why most American urban houses a century ago, even the cramped little row houses that were built for factory workers, had a little plot in back that got at least a few hours of sunlight a day. The kitchen garden and the henhouse went there, and they were as much a part of an ordinary urban household as the pantry.

Thus America a century ago had two separate systems of food production. You would have seen exactly the same thing in most other countries at the same time. If you left your time machine parked in some Iowa barn, hopped the train to New York, and booked passage on a tramp steamer headed around the world, you would find the same double system in most of your ports of call. If you caught the train to Paris while your ship was taking on cargo in Marseille, you would find market gardens around the French capitol using the ancestor of today's deep-bed intensive gardening to keep Paris markets supplied with produce; if you had time to kill in Kowloon while the cargo from Marseille was unloaded, you could travel inland a bit and see the intensive Chinese gardening methods that were also ancestral to today's organic gardening, thriving on little patches of land, while the monotonous green of rice paddies spread in every direction around them.

The great transformation of American agriculture in the middle decades of the twentieth century—exported worldwide under the banner of the "Green Revolution" shortly thereafter—centered on the abandonment of the intensive half of this system and its replacement by extensive farming of the crops that used to be grown intensively. That transformation was only possible because chemical fertilizers could (temporarily) replace the nutrients intensive gardening methods put into the soil, and because petroleum-powered transport could (just as temporarily) make it possible to ship produce across continents and oceans without spoiling, either in processed form or in some semblance of its fresh condition.

The Green Revolution in particular was surrounded by massive propaganda campaigns about feeding the world, but I trust most people by now realize that much of its actual agenda focused on turning the rest of the world into a source of luxury crops for industrial nations. The model they used was the one pioneered in the early 20th century by US fruit companies in Central America, complete with the corporate-backed kleptocracies that contributed the phrase "banana republic" to the English language. The project was a success, in narrow economic terms; the replacement throughout the Third World of small farms growing food for local consumption with big farms growing export crops duly followed, as did the mass expropriation of land that has flooded Third World cities with dispossessed farm families ever since. Recent attempts to turn what foodstuffs are still produced in the Third World into automobile fuel for the industrial nations are simply one more step in the same process.

Unfortunately for the architects and beneficiaries of this system, though perhaps fortunately for the rest of humanity, the whole project depended on huge supplies of fertilizer feedstocks and fossil fuels, neither of which will be available indefinitely. For the world's nonindustrial nations, then, the end of the industrial age thus ushers in a difficult but ultimately positive shift in which the export-driven agriculture of today will come apart at the seams, along with the wild distortions of political power involved in maintaining that system. For the world's industrial nations, on the other hand, the end of a system that kept shoppers supplied with strawberries in January promises to usher in a time of food crisis in which intensive local production will need to be revived in a hurry.

It's thus not accidental that the appropriate tech movement focused on exactly the sort of small-scale intensive organic gardening that is well suited to fill the local-production niche in the human ecology of the future. For that matter, it's not accidental that much of the last half century or so of research and experimentation into organic food growing has focused on intensive production; it doubtless helped that it's a lot easier to afford a backyard or two for experimental garden plots than it is to arrange for 640 acres or so to use some innovative organic farming method or other—though this has also been done, with good results. Some of my readers may be in a position, now or in the future, to try their hand at extensive farming using organic methods to produce grains and dry legumes, and a century from now, maybe half the American population will be making their living that way. They will also have their own kitchen gardens, henhouses, and so on, however, and many more of my readers here and now are in the position to have the same thing. You might be one of them.

The productive potential of intensive gardening, especially under emergency conditions, should not be underestimated. Back in the early 1980s, a team of researchers at pioneering organic-gardening group, Ecology Action, found, on the basis of extensive tests, that it's possible to feed one person year round on a spare but adequate vegetarian diet off less than 1,000 square feet of intensively gardened soil.[8] In the more troubled parts of the future ahead of us, some of us may have to do just that; a great many more of us will need to be able to garden in order to pad out potential irregularities in a food supply that's desperately vulnerable to short-term fluctuations in the price and availability of fertilizer feedstocks and fossil fuels. If this seems reminiscent of the victory gardens of past wars, it should; victory gardens are likely to be a useful template for the survival gardens of the deindustrial future.

A little further down the road, as the resource and energy base for conventional farming begins to run noticeably short, the shift toward a more sustainable extensive agriculture will have to follow. For the time being, though, intensive gardening is the more urgent of the two, and the sooner a significant number of people get to work learning and practicing it, the better.

┌─────────── Exercise for Lesson 7 ───────────┐

It's a common assumption in today's industrial societies that only people living in rural areas have the opportunity to grow any of their own food. In a very few cases, that assumption is justified; in the great majority, it's not. Your assignment for this lesson is to figure out all the places where you could grow at least a small amount of food. If you have a yard, and had the common sense not to settle in one of those grim middle-class ghettos where restrictive covenants tell you what kind of flowers you have to grow on your own property, you're set; if not, you may need to get more creative. If you have an apartment with a balcony or can put out some window boxes in places that will get direct sun, container gardening is an option. Your town may have a neighborhood garden or P-Patch program, or you may have a friend who would happily lend you a corner of his or her yard in exchange for a share of the produce. Make an honest effort to find all the places where you could grow at least a couple of cherry tomato plants or the like.

└──┘

⟫⟫⟫ LESSON 8 ⟪⟪⟪
Understanding Soil

Intensive organic gardening, the keystone of household food production in the appropriate tech movement, can't be approached using the mindset that governs modern conventional farming. That mindset seems reasonable to most people in the modern industrial world, but it really is hard to think of a better example of sheer, rank, bullheaded ecological stupidity.

To grow food crops in today's world, to begin with, do you draw on the dozens of readily available and sustainable flows of plant nutrients, which happen to be the sources that plants have evolved to assimilate most readily? Do you cooperate with the soil's ecology, which has coevolved with plants to store, distribute, and dispense these same nutrients to plants? Do you even recognize that food plants, like every other living thing, are part of ecological communities and thrive best when those communities receive the very modest resources they need to flourish?

Not a chance. No, you get your nutrients from nonrenewable mineral sources, because that's where you can get them in chemically pure forms, even though plants don't get any benefit from chemical purity. You treat soil as though it was a sterile medium serving only as a sponge to hold water and chemicals, and then you do your best to *make* it a sterile medium by pumping it full of chemical poisons so that you can stomp the stuffing out of any attempt by any other living thing to help form an ecosystem involving your plants. Then you wonder why you're stuck in an uphill battle against declining soil fertility, resistant weeds and bugs, water supplies poisoned with chemical runoff, and the rest of it. If some evil genius had set out to invent the most irrational agricultural system possible, I'm not sure he could have done a better job.

Because these are the customs we've grown up with, the ways of thinking fostered by this giddy ecological idiocy seem like common sense. Recent discussions about "peak phosphorus" are a case in point. Our current agriculture, as mentioned in an earlier chapter, relies on mineral phosphates, which are mined from a small number of highly concentrated phosphate rock deposits that are located in odd corners of the world and are being depleted at a rapid pace. The conclusion too often drawn from this is that the world faces mass starvation in the near future—you can't grow food crops without phosphate for fertilizer, and where will we get the phosphate?

There's a point to these worries because our current agricultural system is probably incapable of churning out food at its current pace without those rapidly depleting mineral inputs. Even the rapid expansion of organic farming under way in North America and elsewhere probably won't be fast enough to prevent shortfalls. Still, it has too often been generalized into a claim that the exhaustion of rock phosphate reserves means inevitable mass famine, and this is true only to the extent that current notions of industrial agriculture remain welded into place and nobody gets to work building the next agriculture in the interstices of the present system.

It may already have occurred to my readers, after all, that somehow plants grew all over the world's land surfaces in vast abundance for three quarters of a billion years or so without any chemical fertilizers at all. If this suggests that there's something wrong with the logic that insists that we can't grow plants without chemicals, it should. As proof of this, I'd like to invite you to dust off the time machine we used in the previous lesson and visit a city that doesn't exist anymore, the bustling metropolis of Edo.

It's called Tokyo nowadays, and there's very little left of the city that was there a century and a half ago, but in the Tokugawa era— from 1603 to 1878—Edo, Japan's unofficial capital, had a population that varied between one and one and a half million people. Kyoto, the official capital, and Osaka, the economic hub of the nation, had populations pushing a million each. Even by modern standards, then, these were cities of considerable size, and they were supported by organic rice agriculture that used no chemical inputs at all. The inputs used were human and animal manure, nitrogen extracted from the air by a common and deliberately cultivated species of duckweed, and a great deal of human labor; the outputs kept levels of nutrition in Tokugawa Japan at levels comparable to those of European nations of the same time.

Without rock phosphate, why didn't the Tokugawa-era Japanese all starve to death? Because ours is very nearly the only agricultural system in human history that has ever approached farming with the same idiotic logic Fido chewed up in an earlier lesson: energy and raw materials in one end, products and waste out the other, with no thought as to the long-term availability of the first two or the long-term effects of the latter. Everywhere else in the world, farmers have had a good working grasp of the fact that matter moves in circles. They've long recognized that human beings have to feed the soil if they want the soil to feed them.

In natural ecosystems, in 17th-century Japan, and in every other viable ecology, phosphorus doesn't move in a straight line from phosphate mine to factory to farm to river to dead zone in the Gulf of Mexico. It moves in a circle, from producer to consumer to decomposer and back again, and the soil is the hub around which the entire circle turns. That's true when the producer is grass, the consumer is a rabbit, and the decomposers are soil organisms that process the rabbit's droppings; it's equally true when the producer is a rice paddy, the consumer is a medieval Japanese family, and the decomposers are a different set of microorganisms in a manure heap. Finally, it's just as true when the producers are your garden plants, the consumer is you, and the decomposers are the organisms in your compost pile.

In each case, the result is the same: if you feed your soil, your soil will feed you; if you keep your soil in balance, your soil will by and large keep your garden ecosystem in balance. Learn to think like an ecosystem, and this is obvious; this is among the reasons why the principles in this book's first section need to be studied so closely.

The books listed in the resource section for Part Two of this book—or, really, any halfway decent book on organic gardening— will tell you what you need to know in practical terms to make sense of your soil and work with it. What I would like to communicate here, though, is the sheer wonder of soil. Pound for pound, it's one of the most complex substances in the known universe, a bubbling slumgullion of powdered rock, air, water, bacteria, fungi, other living things, dead plant and animal matter, and humus, which is a jellylike soup of organic compounds that happens to all that dead plant and animal matter, and is complex enough to defy anything but the most exhaustive chemical analysis.[9] By some definitions, soil is a living organism: like other living things, it takes in nutrients, excretes waste, and takes a lively and very active role in the flows of energy, matter, and information that define its environment.

For the organic intensive gardener, certainly, thinking of soil as a living thing is crucial, because it forestalls a major mistake that most farmers and gardeners in the industrial world make as a matter of course. If you try to get milk from a cow or eggs from a hen without feeding it, the results will not be good, and if you try to feed your cow or hen on highly processed industrial products that are convenient for you but not particularly good for the cow or hen, the results will not be much better. Cows and hens have complex nutritional needs that aren't limited to some set of chemicals—cows need roughage, hens need fine gravel for their crops—and if you want to keep them in good health, you need to pay attention to those needs.

Exactly the same principle applies to soil. Like any other living thing, soil has specific food needs that can't be ignored with impunity. You need to pay attention to those requirements because your plants get their food from the soil, and if your soil isn't well fed and healthy, your plants won't be, either. In natural ecosystems, soil gets the food that it needs from dead plant and animal matter. The smart gardener will provide the same in a backyard garden. The ways organic gardeners have learned to do this will be covered in the next lesson.

Exercise for Lesson 8

Learning about soil in the abstract has its value, and you'll benefit from reading about soil in some of the books in the Resources section, but there's also much to be said for getting up close and personal with it. For this exercise, you'll need a few tablespoons of soil that you just dug up from someplace where plants grow, and a magnifying glass; if you have access to a low-power microscope and know how to use it, that gets you extra credit.

Your task is simply to examine the soil as closely as you can. Spread out some of it on a sheet of white paper and go over it bit by bit with the magnifying glass. Notice the shapes and textures of the many different substances that compose it. If the soil is any good at all, expect to find living things—roots, small insects, the tiny smooth-bodied worms that zoologists call nematodes, and more—and the castoff fragments of things no longer living. All these things are part of the underground microlandscape from which garden plants draw most of what they need to live.

If you have the chance, get soils from several different places, including at least one place where very little grows, and compare them under the magnifying glass (and the microscope if you have one). Get to know the difference between fertile soils full of organic material and sterile soils that contain little but rock particles. The clearer your sense of the difference between these two, the better prepared you'll be to build good soil in your own garden.

LESSON 9
Composting and Mulching

Ask a hundred people who don't practice organic gardening what the heart and soul of a successful organic garden is, and you'll get a hundred different answers. Ask a hundred people who do practice organic gardening the same question, and my guess is that a large majority of them will give you one answer: the compost bin. What some of them will go on to tell you, and most of the others know intuitively, is that the humble and lovable compost bin is the template on which the entire structure of any future sustainable society will have to be modeled.

Step out through my back door with me for a moment, and you'll see how this works. My current compost bin is a roughly cubic shape four feet on a side, made of recycled lumber and chicken wire, snugged up to the fence that surrounds my backyard garden. Every day, kitchen scraps and garden waste go into it; every spring, a couple of wheelbarrow loads of rich brown dirt come out of it and get worked into the garden beds. There's lesson number one for a sustainable society: the word "garbage" simply means a resource we aren't clever enough to use yet.

Take a shovel and turn the compost, and I'll introduce you to a few million of my closest friends: the living things that make compost happen. What organisms you get in a compost bin will be determined by how hot and fast you like to do your compost, and this depends on the ingredients you use and how you tend the pile. "Hot" is not a metaphor; a compost bin with the right mix of high-nitrogen and high-carbon materials and just enough moisture can produce so much heat that you'll need to hose it down daily in the summer to keep it from catching on fire. In that kind of heat, very little thrives except the thermophilic bacteria that drive the decay process, but they do thrive. A friend of mine still glows with pride when he recalls the compost pile he built in his 4-H days. It hit a peak temperature of 190°F and finished turning its carefully arranged layers of garden and kitchen waste into ripe compost in only fourteen days.

If you prefer a slower and lazier process, as I do, you can expect to get most of the major animal phyla in your compost bin, along with a bumper crop of fungus and an even larger population of microbes. Most of the critters you can see without magnification will be annelids and arthropods—that is, worms and bugs—and you'll see a lot of them. A good magnifying glass will show you an even more diverse ecosystem. If you have a microscope handy, put a little of the compost in some distilled water, shake thoroughly, pipette a bit of the water into a well slide, and make the acquaintance of a giddy assortment of single-celled organisms.

You also have the option of having a more limited fauna in your compost. People who live in apartments, condominiums, or houses subject to idiotic regulation by homeowner's associations usually find it more functional to use a specialized form of composter called a worm bin. This is exactly what it sounds like, a bin full of dirt that's also full of worms. You feed your vegetable scraps to the worms; they devour the scraps and then excrete some of the best fertilizer you'll find anywhere. Unlike compost bins, worm bins are nearly odorless, easy to run indoors, and can work well on a very small scale. I've known single people living alone who kept worm bins and used the very modest output to keep their potted plants green and growing.

One way or another, the livestock in your compost bin is essential to the composting process because without them, what you get isn't compost, but stinking goo. What happens in a compost bin is exactly what happens in ordinary soil to the vegetable matter that falls onto it in the normal course of nature: decomposers—living things that feed on dead matter—eat it, cycle its nutrients through their own life processes, and then excrete those nutrients in forms that plants can use. What makes a compost bin different is that you, the green wizard, tinker with the conditions so that this natural process can happen as quickly and efficiently as possible so that you can put the results in your garden, which is where you want it. This is where lesson number two for a sustainable society comes in: instead of wasting your time trying to fight nature, figure out what she wants to do and arrange things so that her actions work to your advantage.

Lesson number three requires a little more attention to the details of composting. To keep your livestock happy and healthy, the compost needs to be damp but not soggy, and it needs to get plenty of oxygen. You need to be careful not to overdo the nitrogen—too much freshly cut grass from your lawn, for example, will turn your bin into a soggy mess that smells of ammonia, because grass that's still moist and green has too much nitrogen in it. (Leave it lying for a couple of days before raking it up, so that it wilts and starts to turn brown. Then you can add it to your compost bin with good results.) Different styles of composting, fast or slow, have their own detailed requirements, and worm bins have slightly different requirements of their own.

All these requirements have some wiggle room built into them, but not all that much, and if you stray too far beyond the wiggle room, things won't work right until you fix the problem. Nothing else will do the job. You can't bully or wheedle a compost bin. If you give it what it needs, it will give you what you want, and if you don't, it won't. It really is as simple as that. This can be generalized into lesson number three for a sustainable society: nature doesn't negotiate. If you want her to work with you, you have to give her whatever she wants in return. Oh, and by the way, she won't tell you. You have to figure that part out for yourself, or learn from someone who's already figured it out.

At this point, those of my readers who don't already have compost bins full of a couple of million good friends will have divided into two groups. The first group consists of those people who are eager to get to work making compost. The second consists of those people who are backing nervously away from this book, hoping that annelids, arthropods, and thermophilic bacteria don't crawl through its pages and follow them home. If you're a member of the latter group, you've probably already come up with a hundred plausible explanations why you can't possibly compost your kitchen scraps, or even tuck a worm bin in the utility closet where it will be odorless, harmless, and comfortably out of the way. Still, you know as well as I do that the hundred plausible explanations aren't the real reason you don't want to take up composting. The real reason you don't want to take up composting is the Squick Factor.

The Squick Factor is the ingrained and unreasoning terror of biological existence that's hardwired into the psyches of so many people nowadays. Composting, remember, is about decay. Things put into a compost pile rot, and they get eaten by worms and bugs. Even when you've got your compost in a nice, expensive bin made of textured, recycled black plastic that nobody but a homeowner's association could find objectionable, and even if the only scent that comes off it reminds you of summer meadows from childhood, and even if it can't be smelled at all more than six inches away from the bin, composting triggers the Squick Factor in many people.

There's another name for the Squick Factor: biophobia. Compost is life—damp, oozing, crawling, slithering, breeding, dying and being reborn—and life in the raw scares the bejesus out of most people in the industrial world these days.[10] It's an old, old phobia, but ours is the first civilization in history that has had, however temporarily, enough energy and resources to let its more privileged classes pursue the fantasy of an existence free from biological realities.

The squicky feeling many people get when they contemplate a compost bin is one reflection of our culture's traditional biophobia. If you're going to become a green wizard, though, that attitude is one you're going to have to learn to do without, sooner rather than later, because most of what we'll be doing involves getting elbow deep in life. If the thought of having a compost bin or a worm bin sets off your Squick Factor, it's important to recognize that fact and accept it, but it's also important to go ahead anyway, take the plunge, and discover that the worms in your worm bin are the cleanest, quietest, and least demanding pets you've ever owned.

Composting has another side that isn't always recognized. Even if you've got a thriving organic garden in your backyard, you're probably getting at least some of your food from other sources, and if you're in the very first stages of setting up that soon-to-be-thriving organic garden, you're getting all your food from other sources. The scraps and trimmings that go into your compost bin are inputs to your garden. If you're raking up leaves or mowing your lawn and putting the results into your compost, that's another input. This is a secret function of your compost bin: it's a tool for concentrating nutrients from a wider area into the piece of ground you garden.

More broadly, that's one of the secrets of successful organic gardening: you close up your nutrient cycles as tightly as possible, and you also tap into other nutrient streams that would otherwise become waste and draw them into the eager clutches of your garden. Traditional farmers around the world turned this sort of thing into a fine art, weaving farms and gardens into the wider ecology of the area in richly complex ways. All this needs to be done in ways that don't impair the viability of the systems that provide inputs to your garden, but that can be done easily enough in most cases, given a bit of finesse and a sensitivity to ecological relationships.

Your options here are very broad, and will depend on local conditions. They also extend beyond the practice of composting itself. There are at least three such options that were thoroughly tested in the 1970s, and can be used to add additional organic matter to your soil, alongside what you can get via the compost bin.

The first method is mulching. In many parts of North America, this has become a staple technique of organic gardeners, and for good reason; in other places, for equally good reasons, nobody does it. Back in the day, there were huge debates over mulching; we didn't yet have the internet in the 1970s, but something close to a modern flamewar raged in the letters columns of organic gardening magazine for a good part of the decade before people finally realized that it's a very good technique in some places and a very poor one in others.

The technique is simple enough. You get large quantities of coarse and otherwise unwanted organic material—for example, spoiled hay, autumn leaves, straw, or crushed peanut hulls—and spread a layer several inches thick over your garden beds before planting. When you plant, you clear away the mulch around the seedling or the seed so it can get sunlight. The layer of mulch helps suppress weeds, keeps moisture in the soil, and gradually rots, adding nutrients to your soil.

Mulching can have drawbacks. When I lived in the Pacific Northwest, nobody in their right mind mulched during the growing season because mulch in damp climates is a slug magnet, and slugs in the wet zone west of the Cascades can get up to eight inches long, with appetites to match. I've heard from gardeners who had similar troubles with rats. Of course, you also have to find a source of clean organic matter in bulk, and this can be a challenge in some situations.

The second method is green manure. This amounts to a living mulch for the winter season: something fast-growing that you can sow in your garden beds when the weather starts to cool off, and hoe under in the spring just before planting. The best green manures for small garden use in many cases are clovers, which are legumes and put nitrogen in your soil, and rye grass, which produces a lot of organic matter relatively quickly and breaks down easily in the soil to feed the organisms there. If you mulch, you won't be able to use green manure, and vice versa. Both are good approaches, however, so it's probably worth your while to try them both on different patches of ground to see what works best.

The third method is the tried and true trick of growing plenty of legumes in your garden. Done right, anywhere in the temperate zone, this is a three-step process: you plant peas as early in spring as you can work the soil; you plant beans as soon as the weather is warm enough, and then you plant a second round of peas for fall harvest about the time the summer peaks and begins to decline into autumn. Any kind of pea or bean will do, so choose whatever you like to eat, and plant as many as space permits. If you grow the kind that are eaten green, you can always blanch and freeze anything you can't eat in season, and if you grow the kind that are dried and shelled, an extra pound or two of dried beans or peas in the root cellar is always a good thing to have. Meanwhile, like all legumes, your plants are pumping nitrogen out of the atmosphere and into your soil, where other plants can get at it.

There are many other ways to work the same transfer of nutrients. It's important not to become too dependent on any source of outside nutrients that could be shut off unexpectedly—say, by problems with the economy—and it's as important to make sure that the inputs that you use are the sort of thing that will support the ecology of your garden rather than damaging it, as chemical fertilizers will. Within those limits, there are plenty of options. See what you can come up with.

Exercise for Lesson 9

Your exercise for this lesson is to pay attention to how much organic matter your household produces over the course of an ordinary week. If you already have a composter or a worm bin, your kitchen scraps are already going into a separate container to feed to the critters; if you don't, just for the course of one week, put all the waste vegetable matter from your kitchen and garden into a separate plastic bag. Animal products and oils shouldn't go into the composting process, so leave them out, but any non-oil vegetable product—coffee grounds, tea bags, stale bread, wilted lettuce, you name it—can go in. Get a cheap bathroom scale or some other way of weighing your take, and find out how much you produce. If you don't yet have a composter or a worm bin, all of that could be turning into the best possible soil supplement; if you do, all of that presumably is on its way to your soil as soon as the compost critters get through with it.

LESSON 10
Keeping Them Healthy

There are plenty of good reasons for cycling compost and other good organic nutrient sources into your garden beds, but the most practical of the lot has already been mentioned in an earlier lesson: the health of your plants depends on the health of your soil. This may just be the most important concept in organic gardening, but it's also one of the hardest for most people to grasp, since it flies in the face not only of our culture's attitude toward nature but of our deepest presuppositions about health and illness.

I'm pretty sure that what lies behind American attitudes toward health is a serious overdose of old cowboy movies. The genre of film I have in mind is the one parodied to death on the silver screen by Mel Brooks in *Blazing Saddles*: the honest, happy, and helpless frontier community menaced by evil incarnate in the form of vicious outlaws, corrupt Eastern bankers, local Native American tribes—well, no doubt you know the list at least as well as I do. Along comes a lone gunslinger, with or without a tin star on his vest; plenty of lead gets flung around; when the smoke finally clears, the baddies are sprawled in the dust, the town is saved, and the gunslinger rides off into the sunset to the sound of twangy romantic music.

The story's so hackneyed that not even Hollywood can get away with rehashing it very often these days. Ask most Americans to explain how health and sickness work, though, and dollars to doughnuts you'll get the same basic script, lightly rewritten in medical jargon. The happy, helpless town becomes your body; the black-hatted heavies are germs—that marvelously vague folk label for all the microscopically small living things that may or may not cause human diseases—and the gunslinger is a doctor, blasting away with pills and surgery at the baddies until they all land face down in the middle of Main Street.

If this sounds like an extremely simple-minded way to think about the relationship between a complex system and its environment, that's because it is. Just to start with, dear reader, about ten percent of your body weight consists of germs. You are as much a community as an individual, and many of those germs are essential to your survival; without them, you can't digest food, your immune system starts attacking your own cells, and all sorts of other systems go haywire. Among the trillions of germs that have niches in the ecosystem called "you" are the microbes that cause hundreds of human illnesses—when they get out of hand.

Most of the time, they don't get out of hand because your body is a whole system with an abundance of negative feedback loops holding its bacterial ecosystem in balance. When something upsets that balance, your body has a subsidiary system—the immune system—that normally whacks the offending organism upside the head and gets it to behave. Only when something disturbs the balance of your body so far that the immune system can't stay on top of things do you start getting ill. A brand new organism your immune system doesn't know how to deal with can do the trick, especially if it shows up suddenly and in large enough concentrations to destabilize your body's ecosystem. Much more often, though, it's some other factor that shakes your body's balance and gives the organism a foothold.

That's why the best thing you can do to improve your general health is to eat healthy food, take moderate exercise, and do all those other things you already know about to give your body the reserves of energy it needs to stay in balance. That way, even if you get ill now and then, you can usually count on a less severe case and a quicker recovery, and you won't be flattened by every cold that comes through town. You know that, right? Now apply the same logic to your garden beds, and you'll be well on your way to the organic approach to plant diseases.

Plants don't have the option of picking up and moving to a place that has better soil or the right amount of sunlight. What they get to eat is determined by what the soil is like in the place they're planted. All the other conditions that make them healthy or the reverse are equally fixed once the seed goes into the ground. Your job as a gardener is to know enough about your plants to put the right seed in the right place and to know enough about your soil to put amendments into the soil first. In nearly all cases, the thing you'll need to add to the soil is plenty of compost; good homebrewed compost made out of a mix of kitchen scraps and yard waste contains everything that your soil needs in order to thrive and pass on a suitable mix of nutrients to your plants. Make your soil good and rich in organic matter, and pay close attention to what your plants need in terms of sun, water, drainage, and the like, and you'll solve most plant disease problems before they begin.

Does that mean that plants grown by organic gardeners don't

get plant diseases? Of course not. It means that most of the time, plants grown by organic gardeners get fewer diseases, suffer less from them, and recover more quickly from them than plants grown in depleted soil, even if that soil is pumped full of chemical fertilizers to make up part of the difference. It also means that a great many other challenges faced by gardeners are less of a challenge: healthy soil that contains plenty of organic matter, for example, holds onto water much more effectively than soil that hasn't been enriched by compost and other means, so drought is much less of a problem for organic gardeners. Here again, the parallel with the human body is precise. Everyone gets sick now and again, but if you've made the effort to eat well, exercise moderately, and avoid habits that weaken your immune system, you're going to get sick a lot less often and, most of the time, a lot less severely than people who don't do these things; in the same way, if your garden soil is healthy, the plants you grow there are by and large going to get sick less often and recover more quickly.

You will still have to deal with plant illnesses now and again. Some of them can be tolerated; others can be treated effectively by safe, non-toxic means. In some cases, your best response is to tear up the affected plant as soon as you spot the illness and dispose of it in a way that won't transmit the illness to anything else. Never compost plants that have a serious illness. The compost process will kill whatever's causing the illness, but it may not get rid of spores or other long-lasting forms of the organism. As we discussed in an earlier lesson, when you have some form of matter that's potentially a problem, you need to take it through a long enough loop, with enough transformations, to make it harmless. The loop from plant to compost heap to soil to plant is too short to do the trick.

Choosing or breeding plant varieties suitable for your area, and thus for the specific mix of conditions and plant pathogens that your garden plants will face, is another very important part of an effective response to plant diseases. If you live in a place with hot, wet summers, for example, plants of the cabbage family will be vulnerable to fungal diseases that thrive in such climates. If you've got the fungi in question in the soil, your chances of getting fall cabbages and radishes to survive into the cool weather of autumn may not be good. Different varieties of cabbage and radish, though, have different degrees of resistance to the fungi in question, and if you choose the right varieties, save seed from the plants that thrive best in your garden, and use that for seed stock for next year, you can keep the fungi at bay.

This implies, of course, that you can't simply plant whatever you want and expect it to thrive in your garden. One of the central fantasies behind modern industrial culture is the dream of being able to have whatever you want, wherever and whenever you want it: strawberries at Christmas, pineapple in Alaska, and limitless petroleum reserves in a world that's had more than a century of petroleum exploration all over its surface. It's all driven by the same ultimately rather silly notion that human desires ought to trump the natural order of the cosmos. As a green wizard raising food in an organic garden in harmony with the cycles of nature, you can't afford that sort of arrogance. Part of your work consists of finding out which plants will grow best, with the fewest diseases, in the specific conditions your garden offers. It's only when you've done this that you should be thinking about how to improve those conditions to push the boundaries of what you can grow, should you need to do so.

Exercise for Lesson 10

For this lesson's exercise, you'll need a seed catalog—online or printed, it makes no difference. Flip through it until you find a reference to a plant variety that is resistant to some plant disease, and note down the name of the disease. Then, using the internet, your local public library, the local master gardeners program, or any other resource that comes to hand, look up that disease and learn as much as you can about it. What causes it? What plants does it affect? What makes it more or less likely to damage your garden? What, other than splashing poisons around your garden, can you do to control it? Use several different websites, books, or other information resources to study the disease, and notice whether the sources agree with each other or not.

·✧✦— LESSON 11 —✦✧·
Season Extenders

The distinction between intensive and extensive food plant production has implications that go well beyond the obvious. When you garden a backyard or a few acres intensively, you can spare the time, energy, and resources to do things you can't do on an extensive farm of a few hundred acres, and the payback can be spectacular. One set of these possibilities, to give a simple and practical technology a colorful name, is the art of making and using time machines.

No, I don't mean the kind that can take you to Iowa in 1913 or Edo in 1700, but a subtler and in some ways more interesting variety. The gardener's art requires a close attentiveness to dimensions of time that modern culture doesn't grasp as well as it should. We're so used to thinking of time as an abstract numerical measurement—so many minutes, hours, days, or what have you—that it's often easy to lose track of the fact that for living beings, time always has a qualitative dimension as well as a quantitative one. In the temperate zone, for example, four o'clock in the afternoon is a completely different time for living things in January than it is in August, and twenty days means something completely different for living things at one season than it does at another.

Skilled gardening depends on these qualitative differences. Most of the best gardeners I've known make it a habit to go out into the garden first thing in the morning and stand there, hands in pockets, doing nothing in particular except trying to get a sense of what the garden is doing on that particular day. Most of them also had a collection of ground rules of garden timing—with wiggle room, so they could be adjusted for the vagaries of weather and the like. Choosing the right time to plant particular crops, in particular, is a fine art, but it usually ends up following traditional incantations that are handed down from generation to generation.

In the soggy Western Washington climate where I learned organic gardening, for example, it was the received wisdom that you had to get your peas in the ground by George Washington's birthday in order to get a good crop. Where I now live in the north central Appalachians, I've been told repeatedly by gardeners that it's time to plant corn when the leaves on the oaks are the size of a mouse's ear. I don't know for a fact that there's an Appalachian Standard Mouse whose ears all the local oldtimers have carefully measured, but I'm not sure it would surprise me.

The differences between one time and another are crucial throughout the annual cycle of the garden, but they become especially so in the earlier and later parts of the growing season, when a few degrees of temperature one way or the other can make the difference between successful germination and a failed crop, and the threat of an unexpected frost looms over the garden beds like Godzilla over Tokyo in a classic monster movie. That's when gardeners wish they could somehow conjure up a couple of spare weeks of frost-free weather for spring planting or a spare month of good weather in fall to let some late-ripening crop finish its life cycle.

This is where the time machines come in. Now of, course, we could call them "season extenders" or simply "ways to stretch the number of weeks in which your garden can be productive," but why not go for the more colorful label?

These time machines rely on a principle that we discussed back in Lesson 1 and will be applying repeatedly in the third part of this book—the ability of simple technologies to turn solar energy into useful amounts of diffuse heat. If you've ever climbed into a car that's been left in the sun for a few hours on a hot summer day and yelped when your arm brushed against a vinyl seat heated to the sizzling point, you know the basic trick: a contained space with a transparent cover that lets sun in, but won't let heat out, warms up very effectively in the sun's rays.

There are any number of methods for applying that technique, starting with the cloche. What's a cloche? A transparent, bell-shaped cover with an open bottom that you plump down on top of a plant in spring, before the weather warms. Sunlight streams in through the cloche and warms everything inside—the air, the plant, and the circle of soil within the edges of the cloche—but the heat can't get back out anything like as easily as it gets in. You can spend a lot of money to get elegant glass cloches with or without little vents on top, or you can take ordinary 2-liter bullet bottles, strip off the labels, cut off the bottoms with a sharp knife or a pair of snips, and you're good to go. Cloches are especially useful when setting out seedlings early in the season, when the cold can hinder plant growth and there's still some danger of frost; by the time the plants are well established and starting to bump up against the limits of the cloche, the weather's usually warm enough that you can take them off, give them a good wash, and put them in the basement for next year.

The next step up from the cloche is the row cover. What's a row cover? Imagine a cloche grown long and wide enough to cover a good section of a garden bed—say, eight feet long, two feet wide, and two feet high. Most of the ones I've seen and handled have a framework of wood or one-inch PVC tubing and are covered on the top and sides with clear sheet plastic; duct tape usually plays a role in there somewhere as well. You put it over plants you want to protect against cold and frost, just as you do with a cloche. If you live in a windy area, you'll need stakes to keep it from blowing away; if you live in an area that gets hot sun even in spring, you'll want to make sure things don't get too hot for comfort under the row cover in mid-afternoon. If this becomes an issue, prop it up along one long side to let excess heat get out.

Ready for the next step? That's a cold frame, which might best be described as a permanent row cover. Your standard cold frame has wooden sides and back, and a hinged lid on top, slanted down toward the southern side, that's made of glass or transparent plastic; the front can be wood or glazing, depending on your preference; the bottom is a garden bed. The colder your climate, the more carefully you have to insulate the back and sides and weatherstrip the opening around the lid to get good results. Think of it as a miniature solar greenhouse with access from the top, and you've basically got the idea. Choose the location for your cold frame well, so it will get plenty of sun in winter, and you can get hardy crops from it year round.

The final step, the ultimate backyard garden time machine, is a solar greenhouse. This isn't a simple project, and it needs to be put together by someone with decent carpentry skills. If that's you or someone you know, though, don't hesitate, because a solar greenhouse in a good location can have spectacular payoffs, starting with a year-round vegetable supply. If you can arrange to have it backed up against a south-facing wall of your home, it can turn into a source of solar space heating—we'll be discussing that in a later lesson.

The value of all these methods for extending the growing season is simple enough when you remember that the system that supplies fresh vegetables and other nutrient-rich foods to your local grocery store is utterly dependent on an uninterrupted flow of cheap petroleum-based fuels and agricultural chemicals. It's likely to be a good long while before supplies of bulk grains and dry legumes run short anywhere in North America, but a serious disruption in petroleum supplies—something that could happen for political or economic reasons with essentially no warning—could leave most people in the industrial world scrambling to get access to anything else. Having a thriving backyard garden that keeps you and your family comfortably supplied with vegetables is one kind of security; being able to teach other people in your neighborhood how to do the same thing is another kind of security, and both are worth having.

(Exercise for Lesson 11)

Your exercise for this lesson is simple enough. Sometime toward the middle of the day, when the sun is out, walk around the place where you live, your yard if you have one, or any other place where you've considered the possibility of gardening. Pay attention to where the sunlight falls, and think about the time machines discussed in this lesson: cloches, row covers, cold frames, solar greenhouses. Are there places where you could conveniently put one or another of these? Remember that row covers can go over simple window boxes, and some very small greenhouses are designed to fit in a corner of a patio or an apartment balcony. Even the smallest time machine will give you the chance to learn how season extenders operate; the lessons you learn can be applied on a larger scale when circumstances permit—or demand.

LESSON 12
Saving Seeds

The natural world contains and circulates a vast amount of information. We can read only a very small fraction of it, but that doesn't mean we can do without it. Most of what keeps human beings alive on this planet just now is a function of the information economy of nature.

There are any number of examples, but the one most relevant to the present discussion is the wealth of information to be found in the seeds of any open-pollinated plant. Seeds can be understood in several different ways, but one of the most useful is to think of them as a medium for information storage. Like other media, they will reproduce the information they contain under some specific set of conditions.

You can't look at a DVD sitting all by itself and watch the movie on it; you need appropriate conditions to transform the stored information on the disk into the movie playing on a screen. In the same way, a seed stores information in the form of DNA and transforms it into a plant when you put it in appropriate soil and add sun and water. What makes a seed a far more sophisticated information technology than a DVD, of course, is that when you play a movie, the movie doesn't manufacture new DVDs for all your friends, much less shuffle the movie just a bit every time it plays, and it certainly doesn't do it in a way that tends to produce a better plot and snappier jokes as time goes on.

Seeds, by contrast, do an exact equivalent of this. This is why, when you tap a seed envelope against your hand and send a single seed rolling out onto your palm, you're holding two billion years of stored information. That's how long, according to current paleontology, the process of evolution has been shaping the genetic code of living things. Every generation across that unimaginable length of time has contributed something to the shaping of the little packet of genetic material, nutrition, and protective layers we call a plant seed.

It would be helpful if more people kept this in mind. It would also be helpful if people noticed that the different varieties of any cultivated plant very often contain hundreds or thousands of years of human-derived information on top of the two billion year collection handed over by nature. Instead, this extraordinary wealth is treated even more cavalierly than big US banks treat mortgage paperwork. In particular, most big seed companies sell a handful of bland varieties aimed at the lowest common denominator, producing them each year using hybridizing processes that keep them from breeding true from seed in the next generation. It's very much as though all the more interesting books in the nation's libraries were to be replaced by stacks of identical copies of some gaudy journalistic volume about the pop star du jour.

This would be bad enough if those of us who farm or garden could count on an uninterrupted supply of hybrid seeds for the foreseeable future. For a range of reasons, starting with the end of the age of cheap energy, we can't. The seed industry is one of the world's most monopolistic; the vast majority of all seeds produced and sold in the industrial world come from a tiny handful of vast conglomerates, and their production, transport, and marketing require huge amounts of fossil fuels. It's sobering to realize that the bankruptcy of one big seed corporation just before the Northern Hemisphere's planting season could leave millions of farmers with no access to seed stock. It's even more sobering to consider the far more likely impact of assorted financial and energy bottlenecks on seed availability as an economy in prolonged crisis intersects with spikes in energy, transport, and agricultural chemical costs as industrial society slides down the far end of its trajectory toward history's compost heap.

Fortunately, there are other options. The place to start, unless you happen to have a thriving garden and plenty of heirloom seed stock in place already, is one or more of the seed exchanges and small seed companies that make it their work to keep old, non-hybrid varieties of cultivated plants available. You'll want to look for a source of this kind that carries old varieties from the area where you're living and growing your garden; the varieties that do best in an organic garden in any given area are most often the varieties that did best in that area before chemical agriculture was invented. Finding a relatively local seed source is also helpful. If or, rather, when transport becomes problematic, your chances of being able to access seeds are a good deal better when the source is two counties away than when it's on the far side of a continent.

The ultimate in short supply chains, of course, is to save seeds from the plants you grow. This is definitely a goal for any organic gardener to have in mind, but in my experience, at least, it's not something to take up immediately unless you have no other choice or already have quite a bit of skill in gardening. Every species of plant has its own life cycle, every variety within a species rings its own changes on that life cycle, and the seed—the vehicle by which the plant sends its information on to the future—is often exquisitely sensitive to subtle cues you may not notice if you're not watching. To save seed with any kind of reliable success, you need to know the life cycle, habits, and needs of the plant with which you're working, so that you can give the seed the conditions it needs.

Some of those conditions can be very unexpected to the novice. The seeds of quite a few plants, for example, need to go through a period of fairly intense cold before they will sprout. Keep the seeds of one of these plants in a heated basement and plant them out in spring, and you'll get no crop that year because the seeds are wait-ing for the signal that tells them that winter has come and gone and it's safe to start growing. You can give them the signal they need by storing them in an unheated space or, in the case of seeds that need a good strong chill, sticking them in the refrigerator for a couple of months. Other seeds have a waterproof coating on the surface and may need to be scuffed with fine sandpaper. Still others have unique requirements of their own.

There are also plants that can't be grown from seed at all, or that are best propagated in other ways. Potatoes, for example, don't breed true from seeds—if you let yours go to seed, and they set seed that's viable (which is by no means certain), you're as likely as not to get a half-wild Andean tuber that may not be edible at all. For a good many centuries, instead, potatoes have been propagated by chopping a potato into pieces, each one with its own eye, and burying the chunks in the soil. You need to know that if you want to grow potatoes, and you need to know similar details for a fair number of other plants.

What this means is that you can't save seeds or propagate plants in other ways without learning a fair amount about the plants you grow, which means participating in their ecology and their life cycle. Seeds are not machines; you can't make them do what you want them to do, unless it's what they've spent the last few million years of evolution learning how to do. You have to meet them more than halfway. You can't stand back and wrap yourself in the fraying rags of the failed myth that still convinces too many people that humanity is some kind of alien force, separate from Nature and reduced to poking her with a stick in place of any deeper communion. You have to take your seeds by the hand and join them in the dance.

All our work in this project can be seen as learning how to make sense of the information provided by nature. This is more helpful than restricting our information diet to the far less nourishing fare served up in horse doctor's doses by the various mass production and dissemination schemes of human society. Using solar energy on a small scale, for example, can't be done unless you shake off the conventional wisdom that comes from three centuries of cheap fossil fuels; then you'll understand what you can do with the sun's diffuse heat, and how to orient yourself in space and time to make the best use of it. Choose some other piece of appropriate technology or green living and you'll find yourself in the same place, learning what you need to know from a teacher who doesn't happen to be human and doesn't speak the language you do, but is more than willing to lead you in the dance we call Nature.

Exercise for Lesson 12

Your exercise this time is to locate at least three sources for heirloom and non-hybrid seeds for the region where you live and get access to their seed catalogs, either online or hard copy. If the seed sources you find are nonprofits, send a few dollars their way as a donation to help cover the cost of the catalog—many of them operate on shoestring budgets, and the collapsing economy has left the shoestring even more frayed than usual.

LESSON 13
Breeding New Varieties

Taking part in the dance of nature can go further than saving seed; it can lead the green wizard to take an active role in the process of evolution. I use that word with a great deal of trepidation. There must be other concepts that have been as heavily frosted with myth, nonsense, ideological static, and sheer unadulterated drivel, but I can't think of one. The modern world's obsession with dreams of Utopia has turned Darwin's elegant theory into a shuttlecock batted back and forth by a flurry of ineptly handled intellectual rackets; believers in progress equate it with their imaginary future of infinite improvement; believers in apocalypse treat it as a flat denial of their faith in a future where everyone who disagrees with them will be roadkill on history's highway, and a great many people who don't manage to fall into either camp seem to have lost track of the fact that it means much of anything at all.

My favorite example to date is the woman who put up her hand at the end of a talk of mine in a small and very liberal Left Coast town and said, "But don't you think that children are so much more evolved than adults?" It took a few baffled questions on my part to figure out that by "evolved" she meant "nice," from which I gathered she hadn't spent much time around children lately. When I suggested that she might consider reading Darwin's *The Origin of Species* before using the word "evolution" again, she looked horrified and asked, "Do I have to?"

Evidently so. I don't think it's going too far to ask my readers to take the same advice. Darwin was a very capable writer; *The Voyage of the Beagle* is a classic of travel literature, and his scientific works are written in a more literary style than their modern equivalents. This can take a bit of getting used to, at least for people—the majority these days—who don't read books older than they are. Still, *The Origin of Species* is not an intolerably long book. It can be read in a few evenings, and it's worth investing that much time to watch one of the modern world's greatest intellectual leaps in midflight.

Now it's true, of course, that some of the details of Darwin's theory have had to be changed since his time, as more evidence has come in. This doesn't make *The Origin of Species* any less worth reading. There's much to be learned, in fact, by treating the theory of evolution as an example of the process it describes: the intellectual mutation set in motion by Darwin's work spawned a flurry of variations, which were then sorted out by the selective pressures of further research. As historian of science Thomas Kuhn has pointed out, the same process can be traced all through intellectual history[11]—one piece of evidence among many that evolution, in something like Darwin's sense, is a basic property of all complex systems.

The backyard organic food garden is also a complex system—in fact, it's an ecosystem like any other. Like many ecosystems, it's primarily shaped by the actions of a single species, which in this case happens to be yours. This last point is sometimes exaggerated into the claim that a garden is somehow separate from Nature or wholly subject to human will, which is nonsense. That might be the case if you were growing your plants in a sterile growth medium sealed off from the rest of the world—a bad idea under almost any conditions, and particularly so in an age of diminishing energy and resource availability—but it's certainly not true of the kind of gardens we've been discussing in these pages.

A more useful way to think of a garden, rather, is to compare it to those East African rain forests that are primarily shaped by the presence and activities of elephants, say, or the tallgrass prairies in North America that took their basic ecological beat from the pounding hooves of the buffalo. No boundary separates such ecosystems from the rest of Nature; countless other organisms evolve ways to take advantage of the opportunities opened up by the dominant species, which coevolves in turn to benefit from, or at least not get clobbered by, the other species in the ecosystem. It's not the only way to run an ecosystem—there are plenty of ecosystems that have no single dominant species—but it's tolerably common, and very often highly successful.

That's what a garden is, when it's worked in harmony with local environments and natural cycles; there's nothing uniquely human about it at all. The one wild card is that human beings have pushed the trick of transmitting behavior by learning rather than instinct about as far as it can go, and a lot of our behaviors—in and out of gardens—can change far faster than the glacial pace that natural selection permits. That's our great strength as a species, and it's also our great weakness; what it means in practice is twofold. On the downside, we can never be sure how well our learned behaviors are suited to the demands of our environment. On the upside, if we pay attention to nature, we can pick up useful behaviors in a tiny fraction of the time it would take for those same behaviors to get established as instinct by natural selection.

That's also what a garden is or, rather, what gardening is. Human beings have been a dominant species in a good many environments since we spread beyond Africa most of a million years ago. Despite fashionable claims to the contrary, hunter-gatherer societies manage their environments ruthlessly, using fire and many other tools to encourage natural food production and discourage competition by other living things. What marked the shift from hunter-gatherer to horticultural economies was not a change from blissful dependence on Nature's bounty to brutal manipulation of the Earth, as some simplistic modern ideologies claim, but rather a shift from one mode of ecological management to a more sophisticated one. The key to the latter wasn't planting things where they don't grow naturally—that was already being done in the Paleolithic[12]—but taking a conscious role in the process of coevolution; the new horticultural societies began selecting seeds that had desirable properties and breeding new varieties of plants for their own uses.

Does that sound daunting? If you save the seeds of your own vegetables and replant them in your garden, you're already doing it. Those seeds that don't thrive well enough in your garden's conditions to produce plants healthy enough to set seed themselves are being removed from the gene pool. If you've got the sense the gods gave geese, you're saving seeds from the healthiest and most productive plants, too, which means you've become an agent of natural selection, tilting the playing field in true Darwinian fashion in favor of the most viable variations.

It really is as simple as that. As I mentioned in the last lesson, saving seeds can be a good deal more complicated than it looks at first glance. Breeding new varieties of plants, by contrast, is a good deal less complicated than it looks; once you've worked out the details of saving seeds, all you need to do is pay attention, and you're on your way. You aren't going to create a new species, since you don't have a million years or so to work on it, but you can certainly contribute to the genetic diversity and regional appropriateness of the species you've got.

This can be done equally well with perennial plants, and also with animals. Each of these have wrinkles of their own. Breeding perennials is generally a slow process, though this depends on the plant in question—when you plant an asparagus crown, for example, you normally have to wait for the third season thereafter to harvest asparagus spears and find out how they taste, so breeding a better asparagus is not a project to attempt in a hurry. Animals—at least the small and edible kind that make sense in a backyard garden—breed annually, so developing your own breed is very much an option. You'll need to learn something about genetics, but since most of your rabbits are going into the stew pot before they breed, you've got plenty of opportunities to select the traits you want to develop, and serve up for dinner the ones you don't.

Here, as elsewhere in green wizardry, the key to success is to figure out the way Nature does things, and copy her shamelessly. Imitation is said to be the sincerest form of flattery, but I've come to think that when dealing with Nature, imitation may just be the sincerest form of reverence. Put another way, by learning Nature's ways and adopting them as a basis for our own, we become better able to benefit ourselves and the biosphere at once.

Exercise for Lesson 13

The exercise for this lesson has already been covered: pick up a copy of *The Origin of Species* by Charles Darwin and read it over the next few evenings. This isn't simply so that you can use the word "evolution" in a sentence without unintentionally spouting nonsense, though that does have its charms; no, the point of the exercise is that Darwin studied the methods used by animal breeders in order to grasp the ways that living things can change over time, and his detailed discussion of pigeon breeding is a classic account of how human intervention can shape the evolutionary process. By getting a clear sense of how this works, you'll be much better prepared to do it yourself.

⟭— LESSON 14 —⟬
Wild Helpers

There's a great line in Shakespeare's *Henry IV, Part One* where the English nobleman, Hotspur, and the Welsh wizard, Owen Glendower, are quarreling. "I can call spirits from the vasty deep," Glendower insists, and Hotspur comes right back: "Why, so can I, and so can any man; But do they come when you do call them?" Funny as it is, the quarrel points up one of the things that wizards have traditionally been able to do, and, as a green wizard, you need to be able to do something similar. By the time you've finished with this lesson, you may not be able to call spirits from the vasty deep, but you'll have started learning to call helpful critters from the surrounding ecosystems to help maintain the balance of your garden—and yes, to forestall Hotspur's gibe, they will indeed come when you do call them, if you do it in the right way.

Let's start by reviewing some basic concepts. A garden is an ecosystem managed so that human beings get to eat a significant portion of the net primary production of the plants that grow there. Net primary production? That's the amount of energy each year that the plants in an ecosystem take in from the sun and store in the form of sugars and other compounds that can be eaten by some other living thing. Everything other than plants in any ecosystem gets its fuel from the net primary production of that ecosystem, or of another ecosystem that feeds energy into it.

You're not going to get anything close to a majority of the net primary production of your garden onto your dinner table, by the way, and it's a mistake to try. If you do, you'll starve other living things that depend on a share of net primary production to keep their own dinner tables stocked, and you need these other living things in order to have a healthy and productive garden. Your goal instead is to make sure that as much as possible of the net primary production diverted from your table goes to living things that earn their keep by doing something for your benefit.

Here's an example. A certain amount of each year's net primary production from your garden goes to feed earthworms. Any gardener with the least jot of common sense won't grudge them their share, because earthworms break down organic matter into forms plants can use, and they improve the texture and drainage of soil as they do it. Charles Darwin—yes, the same Charles Darwin who wrote *The Origin of Species*—wrote a brilliant book on the role of earthworms in the creation of topsoil. What he found, to drastically simplify a classic piece of ecological research, is that earthworms are topsoil-making machines, and the more you've got, the better your soil and the higher your crop yields will tend to be.

According to the logic of modern industrial society, you might think that gardeners ought to run out and buy earthworms by the carload, but the logic of modern industrial society is just as wrong here as usual. There are bound to be earthworms in your soil, and since earthworms are hermaphroditic and fertile most of the time, there's generally no shortage of baby earthworms starting out on their slimy and subterranean lives. The question, if you've got a worm shortage, is why so few grow up to become the big pink nightcrawlers that haunt fishermen's dreams.

This is where another of the fundamental principles of ecology comes into play. Liebig's law, named after the 19th-century German agricultural botanist Justus von Liebig, has the interesting distinction of being at one and the same time one of the most consistently valid principles of ecology and one of the most consistently rejected concepts in modern economics. The short form of the law is that for any organism, whatever necessary resource is in shortest supply puts an upper boundary on the carrying capacity of the environment for that organism.

To understand how this works, imagine a plant growing in your garden. That plant has a variety of needs—water, nitrogen, phosphorus, potassium, an assortment of trace elements, and so on. If the soil is short of any one of them, it doesn't matter if all the others are abundant. The nutrient in short supply will determine how well that plant can grow in that garden. All the living things in your garden are subject to Liebig's law, and if you want more of something living in your garden, you need to find out what resource is in shortest supply, and provide it. For earthworms, most often, the sheer amount of organic matter in the soil is the limiting factor, since decaying organic matter is what earthworms eat. Thus, the more organic matter you put into the soil—compost, mulches, green manures, or what have you—the more infant earthworms will mature to massive pink nightcrawlerhood and get to work improving your garden soil.

The same rule governs all the other useful critters you might want to attract to your garden. Bats are a good example. Why so many people fear and dislike bats is beyond me. Any animal that eats its own weight in mosquitoes in a single night ought to be a welcome guest anywhere. Still, the benefits bats bring to the garden outweigh even the simple pleasure of not being eaten alive by the insect world's answer to Count Dracula. Many of the grubs that cause serious damage to food crops—the corn borer, the apple maggot, and more—are the larva of night-flying moths, and night-flying moths are prime bat food.

The limiting resource for bats, far more often than not, is daytime shelter during the non-hibernating months, so one very easy way to bring bats to your garden is to build or buy a bat house and set it in an appropriate place. Both the house and its placement require a certain degree of care, but their preferences are well known.[13] Get a proper bat house in place, and, in most cases, you can count on a crew of bats finding it and taking up residence promptly, and thereafter any problems you may be having with moth larvae will become a good deal less severe.

Birds are the day shift to bats' night shift, and some varieties of birds are well worth attracting to your garden as well. Swallows, swifts, and martins—a closely related group of birds with tapering, pointed wings and a prodigious appetite for insects—are classic examples. Until the advent of chemical agriculture, farmers across North America encouraged barn swallows to set up housekeeping on their farms because swallows do exactly what their name suggests to many daytime insects that make life difficult for crops. Like bats and most species of birds, swallows and their relatives are particular about their homes; here, though, this is a double advantage because homes suited to swallows are uninviting to starlings and other birds that damage crops.

Another set of living things your garden needs is pollinators. The collapse of honeybee populations over much of the industrial world has been all over the news in recent years, and for good reason. Without pollination by insects, many food crops don't produce or reproduce, and honeybees have long been the primary pollinators of most commercially grown fruits and vegetables, with hives being trucked from farm to farm over hundreds of miles in season.

Exactly what is causing Colony Collapse Disorder is uncertain as yet, though a growing body of evidence points to a class of recently introduced and heavily marketed pesticides—neonicotinoids such as imidacloprid—which are highly toxic to bees and can build up to lethal levels in a hive's honey supply. Until the issue gets sorted out, making sure that your garden has backup pollinators is crucial. Domesticated honeybees are one option, but beekeeping is not a project for everyone. Another, far less demanding option is to increase the population of local species of wild bees and other pollinating insects.

Spend some time outdoors watching flowering plants, and you'll quickly discover just how diverse a range of insects can play the pollination game. Many of them are bees of one kind or another, for there are thousands of kinds of bees. Very few of them have the complex social structure and hive life of the honeybee, and even fewer of them have a sting painful to human beings. Most are solitary, harmless, and short-lived, hatching in the spring and mating almost immediately, after which the males die and the females spend the rest of their lives laying eggs in burrows of one kind or another. Each egg will hatch out the next spring as a bee of the next generation, thus completing the cycle. There are things you can do fairly easily to keep this cycle going, and a little research will quickly turn up options appropriate to your area.[14]

Pollinators need something in flower to feed on for the entire period they are active, which varies by species: for orchard mason bees, for example, it extends from March to the end of May in most areas, while for bumblebees, depending on the species, can run from sometime in the late spring well into autumn. The absence of flowering plants can be a limiting factor for all kinds of bees, and if the area around your garden is short on flowers at some point in the season, a flowering shrub or two to fill in the gaps is a good investment. We have a buddleia in our front yard that serves as lunch counter for a dizzying array of daytime insects, including nearly a dozen species of wild bees; your local ecosystem will have appropriate shrubs that will fill the same role.

The same principle can be applied in many other ways. Just as you can encourage a species by figuring out which resource it needs is in shortest supply and providing that resource, you can limit an unwelcome species by figuring out its resource needs and doing your best to make sure that one of those needs is as scarce as possible. As you work with your garden and learn more about the complex ecosystem that an organic garden develops around it, pay attention to places where a little careful tinkering with variables can increase the population of something you want—and decrease the population of something you don't want. It's not so clumsy or random as a pesticide, to borrow and redefine a phrase from *Star Wars's* Obi-Wan Kenobi: an elegant method of the more ecologically sane age toward which, willy-nilly, the pressures of the present are forcing us.

(Exercise for Lesson 14)

Your task for this lesson is to research at least one kind of wild creature—insect, bird, mammal, or what have you—that benefits gardens in your area. The earthworms, bats, swallows, and pollinating insects discussed in the lesson are good examples, but there are many other living things that improve the soil, perform useful services for plants, or devour pests. A little research and a little imagination will turn up plenty of other candidates. When you've chosen your helpful critter, using the internet, the local library, and any other resources that come to mind, try to find out what resources your creature needs in order to survive. Which of those resources might be in short supply in your area? How might you increase the supply of the limited resource, so that you can attract more of your chosen creature to the vicinity of your garden?

✐ LESSON 15 ✎
Home Livestock

When people think about animals in the context of rural homesteading or backyard gardening, odds are the earthworms and bumblebees discussed in the previous lesson won't be the first thing that comes to mind. The reason for this is simple: they just aren't tasty enough. I recall a book I read years ago, with the winsome title *Butterflies In My Stomach: Insects in Human Nutrition*, that made a strong case for dining on insects,[15] but I confess to never having put its recommendations into practice. As for earthworms, I'll leave them to those with bolder palates than mine.

No, the animals most often contemplated in this context are those that provide food a bit more directly, and palatably, for our species. This isn't unreasonable. Though the earthworms, bumble-bees, and other wild creatures that interact with a garden or a farm play a more important role overall in green wizardry, domesticated livestock of various kinds have a crucial place in the backyard food economy. Their task is to take biomass that human beings either can't eat or don't find very nourishing and turn it into more edible and more nourishing forms.

Now, of course, this is not the way modern industrial agriculture generally does things. I commented earlier that if an evil genius set out to design the worst possible way of producing food, his most diabolical contrivances would have a hard time competing with the way we grow food in America today. The animals we raise for human food come out of millions of years of evolution that has fitted them to eat foods that human beings don't, and turn them into foodstuffs of the sort that human beings evolved to eat. Do we feed them their proper foods by putting cows out to pasture, say, or letting chickens scratch for insects and vegetable scraps? Of course not.

Instead, we feed them on grains that could just as well be food for human beings, laced with chemicals and drugs, and "enriched" as often as not with the ground-up bodies of other animals that have been discarded as unfit for human consumption. We do this, mind you, in vast energy-wasting warehouse facilities so overcrowded and poorly managed that the manure, which would otherwise be a valuable resource for improving soil fertility, becomes a massive problem—and, of course, nobody would think of dealing with that problem by any method as sensible as large-scale composting. Meanwhile, the meat, milk, eggs, and other products of this system are a sickly parody of the equivalents that can be gotten from healthy animals fed their natural foods in sanitary and humane conditions.

Plenty of people who object to the appalling conditions and ecological cost of factory farming have responded by swearing off animal foods altogether. This is certainly a choice, but it's far from the only option, and some of the arguments that have been marshaled in defense of it simply don't hold water. Those of my readers who find that a vegetarian or vegan diet suits them should certainly feel free to continue their herbivorous ways, but not everyone finds such diets appropriate to their needs, and those who find a place for animal products on their dinner tables are part of a long hominid tradition. Our australopithecine ancestors ate meat, as indeed chimpanzees do today, and it may be worth noting that no surviving or recorded preindustrial culture anywhere on Earth has had a traditional diet that does entirely without animal products.

It's important to remember, also, that there's plenty of middle ground between eating the products of industrial factory farming, on the one hand, and abandoning animal foods altogether. One way to pursue that middle ground is to buy animal products from local organic ranchers and growers whose operations are open to visits by consumers. Another, though, involves a glance back toward the household economies of an earlier time, when a henhouse in the back garden was as much a part of most urban households as a stove in the kitchen and a roof overhead.

Like food plant growing, in fact, animal raising can be done in one of two ways: extensive or intensive. The extensive approach, in preindustrial societies, is called pastoralism, and it was the foundation of one of the two great human ecologies to evolve out of the hunter-gatherer lifestyle around the end of the last ice age. The other was agriculture, of course, and the two followed parallel paths. Where the early agriculturalists set themselves to domesticate plants they once gathered from the wild, the early pastoralists set themselves to domesticate animals they once hunted. Both new human ecologies had their growing pains and their catastrophic failures, but both worked out most of the bugs and will be as viable after industrialism as they were before it. It's pretty much a foregone conclusion, for example, that the Great Plains four or five centuries from now will be inhabited by pastoral nomads whose raids against the agrarian cultures of the Mississippi basin will impose the same ragged heartbeat on the history of the future as their equivalents on the central Asian plains did for so many centuries in the past.

The cattle herds and nomad raiders of 25th-century Nebraska are a bit too far off for present purposes, though, and the closest modern equivalents are out of reach for anyone who doesn't have enough acreage for the cattle and horses that will define those nomads' lives. This is where the intensive approach comes in. Just as backyard gardens can produce a significant harvest of vegetables when worked intensively, a backyard henhouse or rabbit hutch can produce a steady supply of animal foods when handled in the same efficient and intensive way. This does not mean putting the animals in some small-scale equivalent of a factory farming operation. Instead, it calls for a comfortable shelter and space adequate to the needs of the number of animals you have, along with ample food and clean water, provided by your efforts rather than the less generous habits of nature.

Hens and rabbits are not the only animals that can be raised this way, but for people who don't have enough real estate to set aside a good-sized piece of pasture, they are among the best choices. Both can be kept comfortable and healthy in a relatively small space, thrive on an inexpensive diet, and produce abundantly and reliably if treated well. Hens are particularly good for people with tender feelings toward animals, since you don't have to kill them to be nourished by them. Half a dozen hens will keep a family of humans amply supplied with eggs for most of the year. Rabbits don't have that advantage, and neither do chickens raised for meat; most people I know who raise either one respond to the hard necessity of slaughtering by doing their best to see to it that their animals have only one bad day in their lives.

To be healthy and productive, hens and rabbits need comfortable, well-ventilated, rainproof, and clean housing that is insulated against summer heat and winter cold. In any sort of intensive setting, they won't be able to forage for themselves, so you'll need to keep the feeder stocked, whether it's with food you grow yourself or with something from a local grower or a feed store. They need water, and they need to have their manure hauled away, though they repay this last bit of regular effort by providing some of the world's best raw material for compost. (Animals concentrate nutrients, and a regular dose of chicken or rabbit manure mixed into your kitchen and garden waste in the compost bin will speed the composting process and boost your soil's fertility dramatically.) Animals also need various kinds of incidental care at every stage of their life cycle from birth to stew pot.

What this means, ultimately, is that if you choose to raise hens or rabbits, you or someone you trust will have to be there for them every day of the week, every week of the year. Other animals have different needs, but for all practical purposes, all of them require daily care. The precise requirements are too complex to cover in detail here; they can be learned from the many books available about each animal, and, if at all possible, it is wise to supplement your reading with useful advice from someone who has actually raised the animals in question.

What are some of the other options for small-scale animal raising? Pigeons have been raised for many centuries on a backyard scale. If you have a little more room, ducks, geese, turkeys, and guinea fowl can all be raised successfully. Goats and small pigs are good options if you have more space; the Vietnamese potbellied pigs that were briefly fashionable as pets have gone on to become a staple breed for small-scale pork raising. Backyard beekeeping is also an option, one that has had a recent renaissance in America.

There are more exotic options that can be found with a little searching. Perhaps the most intriguing of the alternatives, though, are fish.

Microscale aquaculture was a central focus of the New Alchemy Institute, one of the most innovative and inspiring of the appropriate tech groups back in the heyday of the movement in the 1970s and 1980s. Tilapia, one of the more popular farmed fish these days, was one of the Alchemists' discoveries; their Arks, or integrated ecoshelters, included tanks for tilapia that provided water and fertilizer, in the form of fish feces to greenhouse crops, as well as a steady harvest of fish. I've never worked with small-scale aquaculture and so have no practical knowledge to offer here, but the concept seems to have worked well in practice. Green wizards who are unfazed by the technical challenges could do worse than look through the papers of the Institute (which are available via several sites online) and start experimenting.

Whether finned, feathered, or furred, animals are a much greater challenge than vegetables. More biologically complex than plants, they are equally more fragile, and they require a great deal more care. The same concentration of nutrients up the food chain that make them so delectable to human beings also make them equally prized by other predators. The sort of hearty nip that most plants can shrug off without incident will put most animals at risk of infection or bleeding to death. Even among green wizards, they aren't a suitable project for everyone, but those who decide that raising small livestock is a challenge they want to take on can contribute mightily to the larders of their households and, on a broader scale, to the resilience of their families and communities in a world where factory farming will be no more than an unhappy memory.

Exercise for Lesson 15

Choose a variety of small livestock that interests you and that you could conceivably raise, either where you live now or in a place where you could reasonably expect to live in the next few years. Using the internet, the local library, and any other resources that come to hand, research their needs for space, housing, food, water, and other necessities. Draw up a tentative plan for setting yourself up with backyard livestock; then spend some time thinking good and hard about whether this is a step that would make sense for you now or in the near future.

LESSON 16
The Unwanted

In the sort of imaginary world where candy canes grow on trees and financial crises caused by too much debt can be solved by adding even more debt, the only animals a backyard gardener would ever have to deal with would either be small livestock who keep the refrigerator full, or helpful critters from the surrounding ecology who come fluttering or slithering in on cue to pollinate plants, turn plant matter into compost, and generally make themselves useful. Alas, we don't live in such a world, and if you have a backyard garden, you'll be dealing with plenty of other animals whose goal in life is to eat the food you grow before you can get to it.

Call them the Unwanted. If that sounds like the title of a second-rate Western movie, that's not wholly inappropriate. Most American gardeners seem to think of them in terms borrowed from cheap Hollywood cowboy flicks: your garden is the inevitable happy but helpless Western town, the animals we're discussing are the black-hatted bandits, and you're the gunslinger with the tin star on your shirt who stands in the middle of Main Street waiting for the baddies to show up, with both hands hovering over the grips of your six-sprayers.

Popular though the image is, it's not a useful approach to managing a garden ecosystem. The idea that you ought to control unwanted animals by squirting poisons over everything may not be the dumbest notion in circulation these days, but it's pretty close. A healthy garden, remember, is one with a diverse population of living things, and the toxic compounds too many gardeners like to spray all over everything are just as deadly to bees and other helpful creatures as they are to the ones you think you need to get rid of. Most of them aren't exactly healthy for you, either, and dumping poisons on your own food supply is not generally considered to be a bright move.

For that matter, it's not even an effective way to get rid of the critters you don't want. It's important to understand why this is the case, because it points up a crucial difference between the idiotic approach that governs so many activities in contemporary industrial society, on the one hand, and the ecological sensibility that ought to guide your work as a green wizard, on the other. Imagine, then, a big field full of a single crop, sprayed regularly with a chemical poison to keep some insect or other from dining on that crop. In ecological terms, what do you have?

What you have is a perfect environment for any insect that can learn to live with the chemical poison. That insect is looking at an abundant food supply, helpfully guarded by a chemical "predator" that will take out other insects who would otherwise compete for the same food supply. Offer evolution a chance like that, and it won't be slow to take you up on the offer—which is why losses to insect pests for most crops in the United States have risen to levels not far below those that were standard before chemical pesticides came on the market, even though most pesticides are being used at or above their maximum safe dosage per acre.

It doesn't help any that nearly all chemical pesticides are single chemical compounds, each of which interferes with the biochemistry of its intended target in one and only one way. The fetish for purity that runs through so much modern technology has many downsides, and this is one of them. Plants that have evolved chemical defenses against insect predators use up to a couple of hundred substances to attack an insect's biochemistry at many different points, an approach that makes it extremely difficult for ordinary random mutations in the insect population to work around them. Rely on a single compound with a single chemical pathway, though, and you make things easy for evolution; one mutation in the right place is all that's needed, and sooner or later the luck of the draw will go in the insect's favor.

The same bad habit lies behind the explosion of antibiotic-resistant bacteria in recent years. Any given antibiotic relies on a single effective substance with a single impact on the bacteria it's supposed to combat; even the sort of antibiotic cocktail used so often nowadays has only two or three active ingredients. Compare that to St. John's wort (*Hypericum perforatum*), which is best known these days as an antidepressant but had a much bigger role in traditional Western herbal medicine as a medicine for wounds. It contains dozens of antibacterial compounds—hypericin, rottlerins, xanthones, procyanins, resins, oils, and more.[16] It's precisely this complexity that makes it impossible for microbes to evolve resistance to herbal treatments.

Crusaders on their way to the Holy Land used to take a wineskin, stuff it with St. John's wort flowers, and fill the skin with olive oil. By the time they got to where the fighting was, the oil was blood red, and they used it to dress sword wounds to keep them from festering in the not exactly sterile conditions of a 12th-century military camp. American laws being what they are, I would probably get in trouble for practicing medicine without a license if I encouraged you to consider herbal remedies such as St. John's wort for infection, or even recommended that you read books on the subject, so of course I'll do no such thing.

The same logic, though, can be applied (with a good deal less risk of legal trouble) to a backyard garden. Instead of trying to get rid of unwanted creatures in your garden by the simple-minded approach of a single poison, you need to change the environment so that it no longer encourages the Unwanted to hang around. That isn't as easy as squirting poison all over everything, since it requires you to learn about the life cycle and environmental needs of each of the creatures you intend to discourage. You'll need to figure out ways to deprive them of things they need, make your garden welcoming to things that eat them, irritate, annoy, and frustrate the living daylights out of them until they throw up their forelimbs in despair and go bother someone else.

Sometimes, a few simple things will do the trick. One classic way to keep raccoons from eating your sweet corn, for example, is to intercrop the corn with some vining plant that will twine all around the cornstalks. Raccoons hate unstable footing, and a tangled, sliding mess of vines will often annoy them enough that they'll settle for the contents of your neighbor's garbage can instead. You can combine that trick with other methods of making life annoying for raccoons. For example, raccoons detest baby powder, and this can be sprinkled liberally on corn ears and leaves to make them leave corn alone. Once the silks have turned brown and pollination is over, you can cover individual ears with old knee-high stockings held on with rubber bands. Do both, and the fellow in the bandit mask isn't likely to bother your corn much.

Some insects, similarly, can be dealt with by the simple expedient of physically removing them. On a large farm this would be a herculean task, to be sure, but in a small intensive garden, it can be workable. Japanese beetles, for example, can be handpicked off your vegetables; do it first thing in the morning, when they're still groggy, and put a tarp on the ground under the vegetables to catch those that fall off. In effect, you become an additional predator of Japanese beetles by putting enough pressure on a population to keep it from getting out of control.

In much the same way, one very effective way to limit the number of slugs in your garden is to find the places they like to hide in the daytime and remove their hiding spots—except for one nice convenient board left flat on the ground in the middle of the garden. Every day, go out and gather up all the slugs that have hidden under the board. You can feed them to your chickens, who will be delighted by the treat. (Gardeners who lack chickens can drop the slugs into a pail of salted water.) Do this regularly, and you can keep the slug population in most gardens down to the point where damage to plants is minor, at best.

Not all problems with the creatures who want to eat your vegetables (or for that matter, your animals) can be solved so easily. Any gardener worth his or her salt has a couple of good books on pest control and takes the time to learn as much as possible about the habits and weaknesses of the insects, mammals and birds that have to be controlled. It's what you have to do if you're going to get food out of your garden. A few good over-the-fence conversations with local gardeners can also clue you in to methods that have been evolved locally. A garden notebook, kept up to date with notes on what works and what doesn't, is another valuable resource.

Perhaps the most important resource, though, is the awareness that in planting and tending a garden, you're working with an ecosystem, not running a machine. Machines require purity, but ecosystems thrive on diversity, which is the opposite of purity. This means that you should have many different crops growing in your garden at any given time, and they should be intercropped rather than grown in nice neat blocks, so that an insect or a plant disease that gets started on one plant can't simply hop to the next one. It means that you should be prepared to use a series of partial deterrents when something that likes to eat your vegetables gets out of balance with the system, rather than attempting a knockout blow that may also knock out something you need.

It also means that you need to accept that a certain number of your plants are going to get sampled by other living things. Concentrate on keeping that number within acceptable limits, rather than trying to get it to zero. You may not want raccoons and slugs in your garden, but they play necessary roles in the wider ecosystem, and, as a green wizard—rather than a poison-toting spray-slinger—your job is to learn to work with the system in ways that work for all concerned.

> ### (Exercise for Lesson 16)
>
> Your exercise for this lesson parallels the ones for Lessons 14 and 15, except the creature you will be researching this time is one of the Unwanted—something that likes to eat the garden produce that you or other people in your area grow. Using the internet, the local library, and any other information resources you have handy, learn about its life cycle and habits and see how many non-toxic ways you can find to limit its depredations.

✎✎✎ LESSON 17 ✎✎✎
Storing the Harvest

Growing some of your own food is a crucial element of green wizardry, but it's not enough by itself. Across the industrial world, people have come to assume that they ought to be able to buy ripe strawberries in December, fresh oysters in May, and any other foods they like, in vast quantity and variety, irrespective of season. That assumption only works if you happen to have vast amounts of cheap energy available. In the real world, foodstuffs show up on their own schedule, not that of the grocery industry. As the age of cheap energy comes to an end, home processing and storage of seasonal foods will become necessary once again, at least for those who don't find scurvy and other dietary deficiency diseases to their taste.

Food storage is a subject that calls up strong emotions, some-times inspiring actions that don't necessarily make much sense. A certain subset of people in the peak oil scene, for example, have become fixated on stashing vast amounts of freeze-dried food in their basements. This is hardly the best option, even if you expect imminent cataclysm. Unless you plan on living out of a backpack during a financial crash, there are many better and cheaper ways to make sure you have some food put by to cope with breaks in the supply chain.

Nor is food storage really about stashing food in a cellar in order to ride out a crisis. A century ago, nearly everybody in America processed food at home for storage, for reasons unrelated to fears of catastrophe. They did it because the foods available year round in a temperate climate don't provide a balanced diet, much less an inviting one. Absent the fossil fuels needed to fly fresh food from around the world to supermarkets in the United States, good sources of vitamin C are mostly to be had in the summer and fall, while meat tends to show up in a lump at slaughtering time in October and November, and so on. If you want these things the rest of the year, and you don't have an industrial economy to take care of that matter for you, you must learn how to prepare foods for storage in season; you must also know how to keep them safely stored until wanted.

The ways that this can be done make a very good lesson in practical ecology. To keep food in edible condition, you have to engage in what ecologists call "competitive exclusion"—that is, you have to prevent other living things from eating it before you do. Your main competitors are bacteria and other microorganisms. You exclude them by changing the habitat provided by the food until it no longer provides the competition with the resources it needs to survive.

You can do that by changing just about every ecological variable you can think of. You can make food too cold for bacteria to survive; that's freezing. You can make food too hot, and keep it in a container that won't let the bacteria back in when it cools down; that's canning. You can make food too dry; that's drying. You can change the chemical balance of food to make it indigestible to bacteria, but not to you; that's salting, brining, smoking, corning, and pickling. You can get sneaky and keep food alive, so that its own immune system will prevent bacteria from getting a foothold; that's root cellaring (there are a variety of other tricks to use with cold-hardy vegetables). You can also get even sneakier and beat the bacteria to the punch by deliberately infecting food with a microorganism of your choice, which will crowd out other microbes and change the food in ways that will leave it in edible condition for you; that's fermentation.

Which of these is the best option? Wrong question. Depending on where you are, what foodstuffs and other resources you have on hand, and how long you expect it to take for various parts of today's society to come unraveled, almost any mix of options might be a good choice. It will likely have to be a mix, because no one method works best for everything. In many cases there's one or another method that's the best or only option.

It's also wise to have a mix because methods of preserving food differ among themselves in another way: some are much more functional in a time of energy shortages than others. If your food storage plans revolve around having a working freezer, you had better hope that the electricity remains on, or you need to make sure you have a backup that will function over the long term—and no, a diesel generator in the basement and a tank of fuel doesn't count, not after the first few weeks of fuel shortage. Blanching and freezing some of your homegrown garden produce is not necessarily a bad idea, but you need to have something in place to power the freezer well before the brownouts start to happen, or you need to be prepared to shift to another preservation method in a hurry, or both.

This points to a second good lesson in practical ecology that can be learned from food storage, though this one's a lesson in practical *human* ecology. Technologies—all technologies, everywhere—vary in their dependence on larger systems. When comparing two technologies that do the same thing, the impact of their relative dependence on different systems needs to be included in the comparison. If technology A and B both provide a given service, and technology A is cheaper, easier, and more effective than technology B under ordinary conditions, technology B can still be the wiser choice if technology A is wholly dependent on an unstable system and technology B lacks that vulnerability.

This much should be obvious, though all too often it isn't. It's embarrassing to see how often a brittle, complex and vulnerable technology dependent on highly questionable systems is touted as better than some simpler, more reliable and more independent equivalent, simply because the former works somewhat better on those occasions when it can be made to work at all. Just as you don't know how to use a tool until you can instantly name three ways to misuse it and three things it can't do at all, it's a waste of time and resources to buy into any technology unless you have a good idea of its vulnerabilities and the ways it tends to fail.

It's worth remembering, furthermore, that there's often more than one way to power the same process. You can dry food, for example, in an electric dehydrator, but in any climate that isn't too humid, you can also dry food in a solar dehydrator. This is basically a black box with small holes in the top and bottom, covered with fine mesh to keep out insects, and trays of screen-door screening stretched on wooden frames inside, with the food spaced on the trays to allow air circulation. The sun heats the box, air flows in through the bottom and carries moisture away through the top, and the food dries with no other source of power. When you've got reliable electricity, an electric dehydrator is more convenient and reliable; when electricity is expensive, intermittent, or not available at all, the solar dehydrator is usually the better plan.

You can also have and use both: the more convenient and reliable technology while you're still on the learning curve (and the larger system that supports it still functions), and the more resilient and independent system, so that you can learn its quirks and be able shift over to it full time when the more complex technology becomes nonfunctional. In the same way, it can make a good deal of sense to blanch and freeze garden produce while you're still learning your way around using home-dried foods, or to can your pickles in a hot water bath while you're still getting the knack of older pickling methods that don't require airtight containers.

In a time of faltering energy supplies, this sort of thinking can be applied very broadly indeed. The strategy of a staged disconnection from failing technologies, made on the basis of local conditions and personal, family, and community needs, offers a pragmatic alternative to the forced choice between dependence on a crumbling industrial system and the unreachable ideal of complete personal or community independence. The backyard garden approach to food is founded on that strategy, and most of the energy conservation and renewable energy methods that are central to the next set of lessons rely on it as well.

There's a reason for this ubiquity: the strategy of staged disconnection is the constructive alternative to collapse. A collapsing society, as it begins to run short of resources, cannibalizes its own assets to replace resource flows and ends up consuming itself. The strategy of staged disconnection avoids this trap; it taps into existing resource flows before shortages become severe and uses them to bridge the gap between existing systems that are likely to fail and enduring systems that have not yet been built. At the same time, if it's done right, it doesn't draw heavily enough on existing systems to cause them to fail before they have to.

That's what could have happened if the industrial world had pursued the promising initiatives of the 1970s, instead of taking a thirty-year vacation from reality that cost us the chance of a smooth transition to a sustainable future. On the collective scale, that's water under the bridge at this point, but it can still be done on the smaller scale of individuals, families, and communities.

Exercise for Lesson 17

This lesson's exercise is simplicity itself: choose a method of food preservation you haven't tried before, and try it. This could mean anything from blanching and freezing some fresh green beans from the grocery store, if you're brand new to food preservation, to your first attempt at home-corned beef or fancy cheese, if canning, pickling, and other methods have been part of your life for years. The project can be as simple or as complex as you choose to make it. The important thing is that you do it, and take that one additional step in the direction of taking charge of your own food supply.

◦⁓ **LESSON 18** *⁓◦*
Using the Harvest

The previous lessons on growing and preserving food ducked one issue that has to be confronted sooner or later: the awkward fact that the food you can produce in your backyard garden does not magically appear in the forms that most Americans are used to consuming. A nation used to eating factory-breaded chicken tenders and Jo-Jos to go is going to face some interesting traumas when food once again consists of live chickens, raw turnips, and fifty-pound sacks of dry navy beans.

It's easy as well as entertaining to poke fun at America along these lines, but the difficulties are very real. A very large portion of today's Americans, provided with a plucked chicken, a market basket of fresh vegetables, and a sack of navy beans, would be completely at a loss if asked to convert them into something tasty and nourishing to eat. The torrent of cheap fossil fuel energy that has so completely transformed the rest of life in the industrial world has worked overtime on America's food system, and this isn't just a matter of how many miles a meal has traveled—how many *factories* has it and its ingredients passed through on the way to your plate?

As the age of cheap energy winds down, it will stop being economically viable to process food in huge centralized facilities and then ship it hundreds of miles in refrigerated trucks to far-flung stores for just-in-time distribution to commuters shopping for dinner on their drive home from work. As that stops being economically viable, those people who know how to produce good meals by some less energy-intensive method will be a lot better off than those who don't. Most people who have had any significant contact with the concept of peak oil will admit this, but all too often a curious thing happens next: they sigh, and talk wistfully about how nice it would be if they had the vast amounts of spare time and the demanding technical skills that cooking meals from scratch requires, but they don't, of course, so it's chicken tenders for dinner again.

You may be thinking something similar, dear reader. You may be thinking that it's all very well to praise home-cooked meals produced from raw materials, but cooking that way is horribly hard work. You've seen the gyrations that actors in chef hats go through in cooking programs on TV, you've glanced over the pages full of exotic ingredients and bizarre processes that make gourmet cookbooks read like tomes of dire enchantment out of bad fantasy fiction, you've seen racks of women's magazines that treat elaborate timewasting exercises disguised as cooking instructions as a goal every family ought to emulate, and you've unconsciously absorbed most of a century of saturation advertising meant to convince you that cooking for yourself from scratch is an exercise in the worst sort of protracted drudgery, and it probably gives you halitosis and ring around the collar to boot, so you really ought to give it up and go buy whatever nice product the nice man from the nice company is trying to sell you.

If all this has convinced you that you don't have time to cook, dear reader, you have been had.

Maybe it's the fact that my grandfather retired after twenty years in the Aberdeen, Washington fire department with a reputation as the best firehouse cook in Grays Harbor County; maybe it's because my stepmother, who taught me how to cook, grew up with the Tokyo working class equivalent of down home cooking in the years during and after the Second World War; or maybe it's because when I left home and settled into my first tiny apartment (two rooms, shared bath), the two cookbooks I had to get me started were the original editions of *Tassajara Cooking* and *The New Cookbook for Poor Poets*. Whatever the reason, the programming somehow failed to stick. I don't particularly enjoy cooking—eating is another matter—and I don't like to spend more than about fifteen minutes of actual work on any meal, but I've always believed in cooking from raw materials, and I've never encountered the least difficulty reconciling those two apparently contradictory things.

In other words, by the time you've gotten off the freeway on the way home from work, fought your way through congested surface streets to the grocery store, found a parking place, done the breast stroke through the crowds between you and the deli counter, caught the attention of a clerk, waited for your order of chicken tenders and Jo-Jos to be heaped into a couple of plastic containers, stood in line again to check out, escaped from the parking lot, fought your way back through those same congested surface streets, and staggered home, I've cooked a homemade meal from scratch and am setting it out on the table. Now, of course, the plum glaze on the pork chops was put up in an orgy of canning two years ago; the vegetable bean soup took ten minutes of knife work and eight hours in a fireless cooker over the weekend and is being parceled out of the fridge a couple of bowls at a time; and it took me a couple of minutes this morning to pick the makings of the salad out of the cold frame. Still, if we count an appropriate portion of those activities as part of my meal preparation time, then we probably also need to count the half hour or so extra you had to work to pay for the difference between the cost of your dinner and the cost of mine.

All this is meant to suggest that there's an entire world of cooking that has nothing to do with elaborate gourmet dishes, on the one hand, or takeout food on the other. A great deal of today's cultural dialogue about food has done its level best to obscure that fact. I have a soft spot for the "Slow Food" movement, but the very choice of that movement's name points out that it will never be anything more than an affectation of the leisured and privileged classes. People who work all day, whether at a job or at home, don't generally have time for slow food, and it doesn't do them any good at all to reinforce a set of assumptions that insist that the only alternative to slow food is the prefabricated industrial product that passes these days for fast food.

What's needed, really, is the revival of the sort of cooking that working class people used to do back in the days before cheap energy made the current food system possible: good food cooked in a way that doesn't place unreasonable demands on the time or energy of people who have many other things to do. The phrase "down home cooking" can be translated into nearly every language on Earth. Every culture uses different raw materials and recipes, so I don't propose to get into specifics here. You probably have a fair idea of the kinds of food you like to eat, and that, rather than random suggestions from archdruids, should be your guide.

One way or another, the sort of cooking I'm discussing will stage a comeback in the age after petroleum. The huge industrial infrastructure that undergirds today's food system is not going to survive the end of the energy surpluses that created it, and when it unravels, people are still going to need to eat. The more people have taken the time to learn the not unduly difficult skills of producing good food quickly, cheaply, and easily, the more time and energy will be available to tackle the many other challenges that we're going to face as the age of cheap energy stumbles toward its end.

Exercise for Lesson 18

Your assignment for this lesson is to take a look at the resources for down home recipes you have available, perhaps in your family, perhaps in your community, perhaps through other channels. The recipes to look for aren't the fancy ones you'll find in glossy recent cookbooks that are meant to gaze scornfully down from the bookshelf and overawe the guests. The recipes you want, rather, are the ones that Grandma Mildred used to make when it was just her and Grandpa George sitting down to dinner on a Monday night, the ones that old Uncle Benny remembers from his days in the merchant marine, or the ones that an elderly lady from your great-grandmother's church wrote out longhand in blue ink to give to your great-grandmother as a wedding gift. You might find them in old mimeographed Grange cookbooks with spiral bindings, or stuffed in the back of the recipe box you got from somebody in the family and never really sorted through, or—well, you get the idea. See what you can find.

RESOURCES FOR PART TWO

There were quite literally thousands of books published during the appropriate tech era on growing, storing, and preparing food at home using low-tech, low-energy and low-cost methods. Two publishing houses—Rodale Press and Garden Way Publishing—published a fair number of the best, but there were also scores of other companies producing good books on these subjects, ranging from big New York firms down to small presses operating out of their owners' basements. The list that follows is thus anything but complete; it includes the books I used (and, in many cases, still use) and can recommend from experience.

It's only fair to point out that there are also very good recent books on many of these subjects, and in some cases—for example, handbooks of plant diseases and insect pests for the home organic gardener—what's available nowadays is far better than anything we had back in the day. You can find these newer books online or in your favorite bookstore.

Abraham, George (Doc) and Katy, *Organic Gardening Under Glass* (Emmaus, PA: Rodale Press, 1975).
A thorough guide to organic gardening methods in greenhouses and cold frames.
Appelhof, Mary, *Worms Eat My Garbage* (Kalamazoo, MI: Flower Press, 1982).

Everybody's favorite book on worm boxes; the revised and expanded 1997 edition is worth getting.

Ball, Jeff, *The Self-Sufficient Suburban Garden* (Emmaus, PA: Rodale Press, 1983).

A methodical, detailed, and practical plan for turning a suburban backyard into the source of most of your family's food in five years.

Bartholomew, Mel, *Square Foot Gardening* (Emmaus, PA: Rodale Press, 1981).

An excellent user-friendly introduction to gardening, particularly good if you've never grown an edible plant before.

Belanger, Jerome D., *The Homesteader's Guide to Raising Small Livestock* (Emmaus, PA: Rodale Press, 1974).

A standard overview back in the day, this provides a good first glance over the options, though it needs to be supplemented with books on the specific animal you have in mind.

Bennett, Bob, *Raising Rabbits the Modern Way* (Charlotte, VT: Garden Way Press, 1975).

The book we used as a guide to rabbit care on the hippie farm where I learned how to raise rabbits. Some of the details are dated but still well worth reading.

Bubel, Mike and Nancy, *Root Cellaring* (Emmaus, PA: Rodale Press, 1979).

A detailed guide to this effective low-tech way to store vegetables and fruit.

Campbell, Stu, *Let It Rot!* (Pownal, VT: Storey Publications, 1975).

——, *The Mulch Book* (Pownal, VT: Storey Publications, 1973).

Two classics on feeding your soil, less finicky than many other books on the subject. Both of them were updated in the 1990s, and the updates are pretty good, too.

Firth, Grace, *Stillroom Cookery* (McLean, VA: EPM Publications, 1977).

A good general overview of old-fashioned food preservation

and preparation methods, ranging from cheesemaking and
beer brewing to curing meats and making pickles.

Fisher, Rick, and Bill Yanda, *The Food and Heat Producing Solar
Greenhouse* (Santa Fe, NM: John Muir Publications, 1976).
The book that made cheap solar greenhouses a standard
element of appropriate tech, a good introduction with plenty
of useful details.

Harrison, John B., *Good Food Naturally* (New Canaan, CT: Keats
Publishing, 1973).
A solid, practical guide to organic vegetable gardening, one of
the standard manuals in the organic scene of the seventies.

Hobson, Phyllis, *Garden Way Publishing's Guide to Drying Food*
(Charlotte, VT: Garden Way Publishing, 1983).
A good introduction to food drying, this was reprinted in
1994 as "Making and Using Dried Foods." Under any name, it's
worth having.

Hupping, Carol, *Stocking Up III* (Emmaus, PA: Rodale Press,
1973).
Repeatedly revised and improved, this hefty volume is still the
gold standard for books on food preservation. If you get one of
the old editions, you may need to update the canning recipes
to meet current standards.

Jeavons, John, *How To Grow More Vegetables* (Berkeley, CA: Ten
Speed Press, 1974).
The raised bed gardening bible, this is still one of the best
books ever written on intensive organic food gardening.
Expect to wear out at least one copy.

Kanable, Ann, *Raising Rabbits* (Emmaus, PA: Rodale Press, 1977).
Another good handbook on rabbits as a backyard meat source,
covering everything from setting up the hutch to tasty rabbit
recipes.

King, F. H., *Farmers of Forty Centuries* (Emmaus, PA: Rodale
Press, 1973 (originally published in 1911).

This was one of the books that everybody who was into organic gardening had on their shelves. A study of sustainable farming techniques in Japan, Korea, and China before the coming of mechanization, it contains more inspiration per page than anything else in the field.

McCullagh, James C., *The Solar Greenhouse Book* (Emmaus, PA: Rodale Press, 1978).

Exhaustively detailed, this covers a wide range of solar green-house techniques and technologies, and includes plenty of examples from the less sunny parts of the United States and Canada.

Mercia, Leonard S., *Raising Poultry the Modern Way* (Charlotte, VT: Garden Way Press, 1975).

Back in the day, everybody I ever met who had chickens in the backyard had this book on their shelves. The updated 1990 edition is worth having.

Minnich, J., *The Rodale Guide to Composting* (Emmaus, PA: Rodale Press, 1979).

Another good reference work on composting techniques.

Newcomb, Duane, *The Apartment Farmer* (Los Angeles: J. P. Tarcher, 1976).

——, *The Postage Stamp Garden Book.* (Los Angeles: J. P. Tarcher, 1975).

Two forgotten classics. *The Postage Stamp Garden Book* is for those who have a little patch of ground to garden, while *The Apartment Farmer* is for those who don't even have that, and still want to garden using pots, window boxes, balconies, patios, and the other limited opportunities of apartment living.

Rodale, Robert, *The Organic Way to Mulching* (Emmaus, PA: Rodale Press, 1972).

A very detailed guide to mulching as part of your organic gardening strategy.

Rogers, Marc, *Growing and Saving Vegetable Seeds* (Charlotte, VT: Garden Way Press, 1978).

A good practical handbook on how to save seeds from your own plants. The updated 1990 edition, retitled Sa*ving Seeds,* i*s* also worth having.

Schuler, Stanley, and Elizabeth Meriwether Schuler, *Preserving the Fruits of the Earth* (New York: Galahad Books, 1973).
A very thorough introduction to food preserving; the canning section needs updating to meet current standards, but the rest of the information remains good.

Seymour, John, *The Self-Sufficient Gardener* (Garden City, NY: Dolphin Books, 1978).
Another classic, lavishly illustrated with water-colored draw-ings, focusing on English methods of small-scale organic gardening.

Stout, Ruth, *How to Have a Green Thumb without an Aching Back* (New York: Exposition Press, 1968).

———, and Richard Clemence, *The Ruth Stout No-Work Garden Book* (Emmaus, PA: Rodale Press, 1971.
Stout introduced mulching to the American gardening scene, and her books on the subject—these two are the best—remain very solid guides to this technique.

Energy

LESSON 19
Using Less Energy

From the thermodynamic viewpoint at the core of green wizardry, all the details of organic gardening covered in Part Two of this book simply deal with one way of turning sunlight into forms of energy that human beings can use. Here in Part Three, we will be discussing other ways of doing the same thing. As with organic gardening, though, certain misconceptions have to be cleared away before it's time to break out the tools and get to work.

The most common of those misconceptions is the notion that the only thing that matters about energy is producing enough of it. This belief deserves to rank among the least helpful legacies of the age of abundance now fading into twilight around us, for two closely linked reasons. The first is that, as fossil fuels go away, our ability to produce what modern people consider "enough energy" is inevitably going to go away as well, and trying to keep that from happening is an exercise in futility. The second is that ignoring the rest of the energy equation—the processes by which we transport, distribute, and consume energy—misses all the options that can make the transition to a post-fossil fuel world easier.

Even among those who think they're being realistic about a low-energy future, the belief that the only possible way to use electricity is to keep uninterrupted power flowing to millions of wall sockets is nearly as sacrosanct as the belief that the only possible way to handle transportation is to find some way to keep tens of millions of cars fueled with as much ethanol, biodiesel, electricity, or what have you, as their drivers can afford. Both beliefs take the temporary habits of an age of excess and treat them as necessities, and both of them box our collective imagination into a futile quest to sustain the unsustainable instead of looking at the real alternative to the extravagant use of fossil fuels.

There is one, you know. People in the contemporary industrial world rarely think about it. Our civilization is so obsessed with delusions of limitlessness that the only alternative to unlimited fossil fuels most of us can imagine is some other energy source that's as least as abundant, convenient, and concentrated. The fact that no other energy source fits these specifications simply adds pathos to the fantasy and motivates the sort of breezy optimism that insists that there must be some vast new source of energy waiting to be found because, basically, we want one, and will whine until we get it. Beyond the daydreams that insist we can replace half a billion years of prehistoric sunlight and geological heat and pressure with our own supposedly limitless cleverness, though, there's a simple, straightforward alternative waiting to take up the slack as fossil fuels go from cheap and abundant to expensive and not nearly abundant enough.

What's the alternative? Using a lot less energy.

The average American uses something like three times as much energy each year as the average European, to support a standard of living that is lower by most of the usual measures. Some of that extravagance is hardwired into the built environment of American society on a scale that individuals can't readily change, but this is far from true of all. Quite a bit is held in place by nothing more than habit and fashion, and can therefore be changed readily, while a good deal more is built into our physical surroundings on a scale that can be changed by individual action.

The sheer wastefulness of today's habits of energy distribution and use is rarely recognized. Behind the wall sockets that produce your electricity, for example, stands the world's largest single infrastructure system, an immense network linking huge power plants and end users via a spiderweb of transmission lines covering whole continents. To keep electricity in those lines, vast amounts of fuel are burnt every day to generate heat, which produces steam, which drives turbines, which turn generators, which put voltage onto the lines; at each of these transformations of energy from one form to another, the laws of thermodynamics take their toll. As a result, only about a third of the potential energy in the fuel finds its way to the wall socket. Similar losses to waste heat also take place when electricity is generated by other means—hydroelectricity, windpower, or what have you—because of parallel limits hardwired into the laws of physics.

When the resulting current comes out of the wall sockets, equivalent losses take place on the other end. Most electricity in today's industrial societies gets turned into light, heat, or motion, and each of these transformations involves plenty of waste. Furthermore, a very large portion of today's uses of electricity are things that could be done just as well without it, with a little ordinary muscle power or some other readily available energy source. That's not even counting the gigawatts that go into lighting, cooling and heating unoccupied rooms, keeping electronic devices on unnecessary standby, drowning out the stars with light pollution, and the like.

Having an energy system geared to so grandiose a degree of excess seemed to make sense in the days when fossil fuels were cheap and abundant. Quite a few absurd things seemed to make sense in those days, and even when they no longer make any sense at all, the habits of that brief interval continue to dominate contemporary thought to an embarrassing degree. Notice how our economists still act as though replacing human labor with fossil fuel-derived energy is always a good idea, even at a time when unemployment is pandemic and the cost of energy is a rising burden on economies around the world.

The difficulty here is that most current conversations about the future of energy are trying to figure out an answer without first making sure that what's being asked is the right question. "How can we keep ourselves supplied with cheap abundant energy?" is the obvious question, and it's also the wrong one. The right question—the question that we should be asking—is something more like "How much energy will we have in a future after fossil fuels, and what are the best ways to produce, distribute, and use it?" That question was discussed during the heyday of appropriate tech, and the answers that emerged from that discussion do not support the notion that the way we happen to do things nowadays is necessarily the best way to do things.

There are many reasons for thinking, in fact, that trying to maintain a large-scale electrical grid in a future of scarce energy is a fool's game. To run a large-scale grid, you need to be able to produce huge amounts of power every second of every day. It's hard to get that much power that reliably by any means other than burning a lot of fossil fuels, or by pursuing other gargantuan power systems—for example, nuclear power plants or giant wind turbines—that can only be built and maintained by a society with ample fossil fuels to run a transportation network, power machinery to mine rare minerals, and so on. These won't be viable in a future where fossil fuels and other forms of concentrated energy are in very short supply.

Still, this is less of a crisis than it seems. Gigawatts of power are necessary for a large-scale power grid. They aren't necessary for the homes and small businesses that make up the great majority of end users. Get rid of the pointless excess that dominates our current approach to energy, and limit electricity use to the things it actually does better than other energy sources, and a 12-volt circuit at very modest wattage is very often all you need. Powering a 12-volt circuit at modest wattage is child's play, and can be done by any of a baker's dozen or so of readily accessible technologies that can be built, maintained, and used by any moderately skilled handyperson. If you adjust your lifestyle to reality rather than expecting reality to adjust itself to your lifestyle, furthermore, living with an intermittent electrical supply simply isn't that difficult. After all, it was standard practice in our great-grandparents' time.

Eighty years ago, radio was the hot new communications technology, and even in farm country, where the electrical grid took its sweet time to arrive, most families who could scrape together the cash had a radio in the living room. Where did they get the power? Batteries, two of them per radio: an A battery to heat the filaments on the vacuum tubes and a B battery to provide the working current. The A battery needed frequent recharging, and wind power was among the standard options for that; the iconic windmills that dotted rural America three quarters of a century ago had plenty of uses, and as often as not, recharging radio batteries was one of them.

Long before electrical grids extended out of America's urban cores, in fact, it was fairly common for households elsewhere in the country to have a modest amount of electricity to hand. That was used for the handful of things available in those days that electricity powered more efficiently than any other form of energy— for example, electronic devices such as radios and phonographs; safe, smokeless lighting for the parlor and the kitchen for a few hours after sunset. (Nowadays a well-insulated refrigerator and the pump for a closed-loop active solar water heater would be worth adding to the list.) Those things that electricity does inefficiently and other energy sources do well—for example, providing diffuse heat or high-torque mechanical energy—people did by other means. Fairly often, a certain amount of muscle power was required, but that's an inevitable reality of life in a world after abundance.

The same principle applies across the entire range of green wizardry. Sunlight can be turned into electricity on a small scale easily enough, but there are always huge losses to waste heat. On the other hand, if what you want is diffuse heat—the sort of heat that will warm a room, heat a bath, or bake a loaf of bread in a solar oven—solar energy is exactly what you want, because it will do this with almost no waste at all. When planning for solar energy, in other words, it's best to do as much as possible with the diffuse heat sunlight provides so readily; sunlight should be converted to electricity only when electricity is the only thing, or the best thing, for the job.

Apply the same logic across the board, and you end up with the most probable energy system of a world after abundance: a patchwork of different energy sources and applications, right down to the level of the individual household or business. In North American households three quarters of a century ago, that was standard: the radio ran off the A and B batteries, the stove was fueled by wood from the woodlot, two lamps in the parlor ran off batteries charged by the windmill while the rest burnt kerosene, the sewing machine had a foot-operated treadle, the water was pumped by the windmill and heated by the sun—solar water heaters were hugely popular in the 1920s, especially but not only in the Sun Belt.

One consequence of this crazy quilt of energy options is that if something disrupted access to any one source of energy, the rest of the household chugged on unaffected. Compare that to the electricity-dependent household of the present, where a blackout renders the whole household nonfunctional. This is just one way in which expectations formed by the extravagance of the recent past are not a useful guide to the best options available to us in the post-peak future.

┌─────────────────────────────────────┐
(**Exercise for Lesson 19**)

Get a sheet of paper and a pen or pencil, and go through your house, making a list of every appliance that's powered by electricity. Once you have your list, go over it again and make a note next to each appliance: *heat* if the appliance is meant to produce heat, *cold* if the appliance is meant to produce cold, *motion* if the appliance is meant to move something, *light* if it's meant to produce light, *information* if it's meant to convey or process information. When you've finished, go over the appliances again and try to find at least one way of providing an approximation of the same service that uses no electricity at all. Write it down next to each entry. Don't worry at this point about whether you could use these alternatives in your present lifestyle; the point is to get you thinking about alternatives, and free up your imagination from the straitjacket of abundant energy.
└─────────────────────────────────────┘

⚞ LESSON 20 ⚟
Caulking and Weatherstripping

Diffuse heat, as suggested in the previous lesson, is a very useful thing. Diffuse heat is what you need to heat your home, your bath and your dishwater, to cook your meals, and to do most of the other things that make for a comfortable and healthy existence. It is also the easiest kind of energy to supply, because all other kinds of energy turn into diffuse heat whenever they can. Turning sunlight into diffuse heat is easier and less wasteful than turning it into any other kind of energy, so most of the applications of solar energy that were worked out during the heyday of appropriate tech focus on solar heating of one kind or another.

Since space heating and hot water make up a very large fraction of the energy consumption in the average American household, these are obvious places to start. Still, it's not enough to slap a solar water heater and some thermosiphoning air panels onto your shoddily built modern McMansion and call it good. Diffuse heat has its drawbacks as well as its strengths, most of them are a function of its diffuseness. The work that a given quantity of energy can do, remember, depends on the difference in concentration between the energy and its surroundings. When you've got plenty of energy sources that will produce highly concentrated heat, you can afford to be sloppy about making the most of energy, but when all you have is the diffuse heat in sunlight, you don't have that luxury. You have to make your energy work for you.

The standard slogan back in the day, along these lines, was "weatherize before you solarize." What this means is simply that making your home, or anything else, conserve energy effectively is the critical first step that needs to be taken before you replace highly concentrated fossil fuels with relatively diffuse renewable energy sources, such as sunlight. In order to do this, you need to know how heat leaves the places where you want it, and this requires a good basic knowledge of practical thermodynamics.

You already know that energy always moves from higher concentrations to lower concentrations. Translated into terms of heat, that means that heat always passes from a hotter body to a cooler body, never the other way around. Any basic physics textbook will tell you that it can pass in three ways: conduction, convection, and radiation. Conduction is the movement of heat through direct contact—think of the way a metal spoon left in a saucepan of hot soup on the stove picks up heat from one end and transfers it to the other: that's conduction. Convection is the movement of heat through currents in air and water—think of the way the soup in a pan on the stove moves. As the soup contacts the hot metal on the bottom of the pan, it rises, and the soup in contact with the relatively cold air sinks: that's convection. Radiation is the movement of heat through photons, mostly vibrating in the infrared part of the spectrum—think of the way your hand feels heat when you hold it a couple of inches from the side of a pan of hot soup: that's radiation.

Those are the three classic ways that heat moves. What your introductory physics book may not tell you, though, is that there's a fourth way, a hybrid of conduction and convection, that is responsible for up to half of the heat loss in an average North American house. This is infiltration: cold air from outside leaking indoors. Technically speaking, infiltration is balanced by exfiltration, which is the process by which the nice warm air in your house goes outdoors so it can radiate its heat to the environment; in practice, since infiltration and exfiltration use the same kinds of leaks and can be fixed in the same ways, the label "infiltration" works for both.

What makes infiltration the best possible starting point for energy conservation is that it's far and away the easiest and cheapest source of heat loss to fix. The gear you'll need are a caulk gun, several (usually, quite a few) tubes of good weatherproof caulk, and an assortment of weatherstripping supplies for doors and windows. You'll also need a sturdy scrubbing brush, cleaning supplies, and a pair of gloves you don't mind ruining. Your local hardware store will provide you with everything you need.

If you've never used caulk or a caulk gun before, you can find good instructions in any decent book on home repair. Your goal is to find all the little cracks where air is leaking into and out of your home, so you can seal them with caulk. There are almost certainly a lot of them: along the baseplate where your house joins its foundations, along the frames of windows and doors, in the little holes drilled through the walls by the guy who installed cable television or internet service, around outdoor water faucets, and the list goes on. Search the inside and outside of your exterior walls and find every crack and gap; make sure the surfaces are clean, so that caulk will stick to them, and then, to borrow a phrase from one of my instructors, caulk those puppies.

Now of course you're not going to caulk the moving parts of your windows and doors, since you need to be able to open and close them. (Nonmoving parts of windows can and should be caulked; if the windows are old, they probably leak like sieves.) For doors and windows that open, you need weatherstripping. There's a dizzying range of products available, most of them haven't changed much since I studied this stuff in the 1980s—for that matter, most of them haven't changed much since the 1950s and 1960s-era home handyman books I collect started to include chirpy little articles titled "Saving Money with Weatherstripping!"—but different door and window situations call for different kinds of weatherstripping, so take your time and explore your options. Books on home repair will often give you plenty of information, and if the staff at your local hardware store know their job, they can advise you as well.

A few other details can help you close off other air leaks. Electric sockets and switches on the inside of exterior walls are often the places where air leaks get into your living spaces; your hardware store will sell you inexpensive foam gaskets that go behind the faceplates to take care of this. The hatch into your attic, if you have one, needs to be weatherstripped, since your attic is probably vented to outside (if it isn't, it should be; more on this in a later lesson) and can leak a lot of heat. Finally, if you've got an open fireplace, one heck of a lot of warm air is rising out through the chimney to heat the great outdoors. A set of glass doors or some other way of closing up the fireplace opening when it's not in use will be well worth your while.

By the time you finish caulking, weatherstripping, putting in foam gaskets, and installing glass doors on your fireplace, you may be wondering how any air is going to get into your house so you can breathe. With the relatively simple technologies we're discussing, that's not a problem; if you do a good job, you're probably going to be able to reduce the rate at which air flows through your house by about half, which means that you're going to save about half the money that infiltration currently costs you—roughly ten to twenty-five percent of your heating bill, in other words—without causing any problems worth noticing. It's possible to seal your home so tightly that you have to start worrying about indoor air quality, but that requires much more complex methods.

Back in the day, there was quite a bit of talk about superinsulated houses that used massive insulation and very thorough air sealing; more recently, the Passivhaus system, which was invented in Germany and has recently taken root on this side of the Atlantic, uses similar tools to get even more remarkable effects. A good superinsulated or Passivhaus home will keep itself comfortable year round in most climates with no furnace or other heating system at all. These aren't things you can do yourself—you'll need to hire a professional—and you're going to shell out quite a bit of money to do it, but if you've got the funds to invest, free heat for life is a pretty good payback.

On the other hand, most people who are interested in energy conservation these days don't have the kind of spare income that would allow them to drop five figures on a Passivhaus remodel, even fewer will have that kind of money as the economic unraveling of our society picks up speed, and it's exactly those among us who don't have the funds to spare for that sort of project that have the most urgent need to save money and energy just now. The sort of low-tech, low-cost forms of energy conservation discussed in this lesson are much better suited to green wizards on a budget, and they can knock a considerable chunk off your heating and cooling bills immediately—leaving you with money you can save up for more elaborate projects later on.

Exercise for Lesson 20

This exercise is best done on a windy day. Get a candle, light it, and go around to the windows and doors of the place where you live. Being careful to keep the candle flame away from anything flammable, put it near the places where air might be leaking in from outside—the cracks around the bottom, sides, and top of exterior doors, the edges of windowpanes, electrical sockets on outside walls, holes drilled in outside walls to allow television cable or internet into the house, and so on. If you have a basement, go along the join between the concrete and the above-ground walls and see whether there are gaps. The candle flame will show you: in still air, it will rise straight up, but any breath of air movement will send it dancing to one side, and a really good leak can blow it out. When you've finished the exercise, think about the fact that for every molecule of cold air being blown in by the wind, there's another molecule of air that you paid good money to heat being blown out a crack somewhere else to warm the great outdoors.

⟡— LESSON 21 —⟡
Insulation

Important as it is, infiltration is rarely the largest source of heat loss in a house—just the easiest to fix. The most important are the three classical methods of heat movement—conduction, convection, and radiation—which allow heat to slip out through the walls, floors, and ceilings of your house even when you've caulked and weatherstripped every crack and cranny in the place. The next step in slowing down the movement of heat out of your home in cold weather, and into your home in hot weather, is insulation.

We can start with the fact that most North American homes leak heat like sieves. Part of that, as discussed in the previous lesson, is a function of the fact that as cold air flows into your house and warm air flows out of it, a portion of the heat you paid for ends up warming the great outdoors by some tiny fraction of a degree. Part of it, though, is a function of the fact that the ceilings, walls, and floors of most North American homes offer inadequate resistance to the flow of heat. Put your hand flat against the inner surface of one of your home's exterior walls some cold winter night; if the wall feels colder than the inside air—and in most houses and apartments these days, it will—you're feeling the result of low resistance to heat flow.

Resistance to heat flow is measured by what, usefully enough, are called R-values. Every material has its own R-value, and the R-value of most construction materials isn't very high—a half-inch thick sheet of plywood, for example, has an R-value around 0.62, which is actually less than the R-value of the thin layer of air that clings to the inside surface of each wall of your house. A wall of standard frame construction, without insulation, has an R-value somewhere around 4.25. Insulate the same wall with a roll of standard glass-fiber wall insulation, and its R-value goes up to an average around 12.75, which means that heat takes three times as long to flow through it.

Putting insulation inside a wall can be a complicated operation if the wall is already built, I've heard very mixed reports on the kinds of insulation that are pumped into an already-built wall through holes drilled for the purpose. Since this book focuses on retrofitting and other measures on a budget, you'll want to consider insulating your walls only if you're already planning on ripping out the drywall or replacing the siding—there are rigid-board insulation products you can put on the outside of your house, between the old sheathing and the new siding, with good results. You can also borrow a trick from the Middle Ages and use fabric hangings of various kinds to insulate your walls; we'll discuss those in more detail in the next lesson.

If you don't have major domestic surgery or fabric hangings in mind, though, your best options lie elsewhere. The first is right above your head. Heat rises, of course, and so most houses lose a substantial portion of their heat through the ceiling and roof; walk around the neighborhood on a day after there's been light snow, an inch or so, and you can often see a dramatic difference between the parts of roofs that have heated space beneath them and the parts that don't, evidence of the amount of heat rising straight up through the roofs. In most cases, the best place to put your insulation is the floor of the attic, right above the ceiling. You want plenty of insulation there, too: an R-value of 60 in ceiling

insulation is not excessive in any place with cold winters, hot summers, or both.

The reason you want to have the insulation just above the ceiling is threefold. First, since heat rises, you want as much as possible
of it to stop rising while it's still in your living space, rather than
warming the cobwebs in the attic; second, your attic can then be
vented, and if you live in a snowy area this will keep the roof cold
and prevent the freeze-thaw cycles that generate ice dams along
the eaves and potential repair bills in three or four figures; third, if
you live in an area with hot summers, a vented attic means that the
solar heat that builds up under your roof can be vented out into the
outside. Good insulation keeps it from trickling down and making
your living space miserable in August.

If you have an unheated air space underneath the first floor,
whether it's a basement or a crawlspace, that's another good area to
insulate. Here your insulation should go right up under the floor,
and you normally do this from underneath. Insulation with an R-
factor of 19 or so is usually enough to keep the cold from creeping
in. In most cases, if you caulk, weatherstrip, and insulate above the
ceiling and under the floor, you'll cut your annual heating bills by
up to half—using your own unskilled labor and supplies your local
hardware store will be happy to sell you. The one further step to
add to the package is insulated window coverings, which we'll be
discussing in the next lesson; they provide a huge gain, since your
ordinary single-pane window has an R-value around 1, but they're
complex enough to require a discussion of their own.

There's a complicating factor with insulation, though, and that's
the effect of water vapor. The amount of water vapor that can be
carried by a given amount of air depends on the temperature of
the air; as the temperature drops, the vapor turns back into liquid
water and condenses onto any available surface. That's where the
drops on the outside of a bottle of cold beer come from: the air
close to the bottle becomes chilled, and the water vapor contained
in the air condenses out onto the glass. When it's cold outside and

warm indoors, you get a similar temperature drop in the middle of an insulated wall, and any water vapor that's present will condense out onto your insulation. Water conducts heat much better than insulation does, so condensed water causes a sharp decrease in the R-value of your insulation, and it can also lead to problems with mildew and dry rot if it goes on long enough.

You prevent this with vapor barriers. The standard vapor barrier these days, I'm sorry to say, is plastic sheeting, which is waterproof, moldproof, airtight, and cheap; finding a way to make something comparable from renewable materials would be a very worthwhile project. In the meantime, though, you need a vapor barrier if you're going to insulate, and that means plastic sheeting, well overlapped and taped with duct tape. A crucial point you need to remember is that the vapor barrier always goes on the side of the insulation that's going to be warmer during the heating season: in other words, when you're insulating the ceiling from the attic side, or the floor from the basement or crawlspace, the barrier goes in first and the insulation afterwards, but if you've torn out some drywall and are installing insulation, the insulation goes in first and the vapor barrier goes on over it, right beneath the new drywall.

Here again, if you have the money to spend, you can insulate the bejesus out of a house and get it to the point where body heat, cooking, and a bit of additional boost on really cold days will take care of your heating needs. The Passivhaus system mentioned in the previous lesson uses this as one of its ingredients, and it's certainly an option, but it's going to set you back a chunk of change. Those who don't have that kind of money to spare—and they will likely be the very large majority over the decades to come—can still keep themselves, their families, and anybody willing to learn from them comfortably warm in winter on much less energy by using the simpler methods discussed here. Further on, as current housing stock wears out, other techniques will need to evolve; with any luck, enough knowledge of how insulation works can be passed on through the upcoming crises to make that process easier than it would otherwise be.

Exercise for Lesson 21

Your exercise for this lesson is simple: figure out, as best you can, how much insulation you have in the place where you live. If you live in a house, that means getting into the attic and the basement or crawlspace with a ruler, and measuring the thickness of any insulation you find there; if it's the kind that comes in rolls with a kraft paper backing and has its R-value printed on it, that's good for extra credit. The walls will give you more trouble; if you're lucky, you might be able to unscrew the cover plate on an electric socket on an exterior wall and get a glimpse past the junction box at the insulation, or lack of it, between the drywall and whatever you have underneath your siding. If you live in an apartment, you may not be able to find out—try telling the manager that you need the information for a class you're taking—but give it a shot. If worse comes to worst, putting your hand flat against an exterior wall on a cold night will give you some idea of how much heat you're losing.

LESSON 22
Window Coverings

If you've caulked and weatherstripped your home, and have a decently thick layer of insulation in the attic, your windows are where the largest fraction of your remaining heating bills go dancing out into the great outdoors. Window glass has an R-value (R means resistance to heat flow, remember?) right around 1 per layer of glass, so a double-pane window has an R-value of 2, or maybe a bit more: that is to say, not much. Interestingly, this is true no matter how fancy or expensive the windows happen to be: you get an R-value of 2 or so from an old-fashioned single-pane window with storm windows slapped on the outside, and you also get an R-value right around 2 from a very expensive vinyl-framed double-pane window with the space between the panes pumped full of inert gas, or what have you. If you want a higher R-value, glass is not going to give it to you.

One point worth taking home from this last comment is that if you've got windows that don't serve a useful purpose, getting rid of them, permanently or temporarily, may be your best option. It takes a certain amount of skill at carpentry to take out a window and seal up the opening so that the resulting wall is weathertight and well insulated; if you don't happen to have the skills, your local handyperson can do the job in a day or so, and it's often money well worth spending.

If you don't feel confident in doing anything so drastic, get some rigid-board insulation from your nearest lumber store, cut it to fit exactly into the window opening from inside, and then cut a sheet of hardboard to fit the same opening, inside the insulation; glue the insulation to the hardboard, paint the hardboard to match the wall, weatherstrip the edges of the hardboard so that you've got a good tight seal around the sides, top, and bottom to prevent air leaks, slide it into place and you're good to go. If you live in a place with cold winters, closing up half a dozen windows in this way during the cold season can save you quite a bit on your heating bills. If you live in a place with hot summers, equally, sealing up a couple of windows facing south or southwest can keep your place much cooler.

What if you want something more easily movable, so you can catch the rays of the winter sun when it's out but close things up easily at night, or screen out the summer sun by day but get that nice cool breeze once the sun goes down? Here we come to one of the great forgotten secrets of the Seventies appropriate-tech movement, the fine art of insulated window coverings.

I had the chance to learn about those personally in my teen years. In 1977, my family moved from a rental house in a down-at-heels Seattle suburb to a larger and more comfortable place we actually owned—well, subject to mortgage and all that, but you get the idea. The one drawback was that the new place was expensive to heat, mostly because most of the main floor's walls facing southeast, toward a stunning view of the Cascade Mountains, consisted of single-pane windows. Insulated window coverings were much talked about in those days of high energy costs and state-funded conservation programs; my stepmother found a pattern, fired up her sewing machine, and made what amounted to a set of inexpensive quilts—faced inside and out with the ornately printed sheets popular in those days, and filled with polyester batting—rigged to slide up and down like Roman blinds. They went up in the morning and down with the sun, and the monthly heating bills dropped by a very noticeable amount.

There are dozens of designs for insulated window coverings— or, more precisely, there *were* dozens of designs. It will take you a bit of searching to find them nowadays, as a result of the thirty-year vacation from reality American society took after 1980. All the designs have certain things in common. The first, obviously enough, is that they put a bunch of additional insulation over the window. How much? A good rule of thumb is that your windows, with window coverings in place, should be as well insulated as the wall on either side—for an uninsulated wall of normal American housing construction, this means around R-5, and up from there as your level of insulation improves.

The second common feature is that the window covering should be sealed around the window, especially at top and bottom. Conventional curtains, open at top and bottom, can actually increase heat movement by convection: in cold weather, air next to the window glass is chilled and flows out the bottom opening, making a draft across the floor, while warm air gets drawn in through the bottom opening and flows across the glass, cooling as it goes. Stop that "flue effect" and you instantly make the room more comfortable. The insulated shades my stepmother made were pressed right up against the wall above the windows; they had little magnets sewn in along the edges to hold them against metal strips in the wall beside and below the windows; there were many other tricks used to do the same thing.

The third common feature is that the window covering should contain a vapor barrier. Ours didn't, which meant that the windows were thick with condensation when the shades went up in the morning, so they often had to be mopped off with a rag. A layer of something waterproof, on the side of the insulation closest to the interior space, will prevent that, and avoid problems with mold, water damage, and the like.

Beyond these three points, the options are nearly unlimited. It's entirely possible to use something like ordinary curtains to get the same effect, as long as they have something holding them tight against the wall on all sides of the window opening. Shades were a very common approach, and so were shutters of various kinds, hinged or sliding or even concealed within pockets built out from the walls. One of the most elegant examples I know involved built-in bookcases along a northern wall; there was a gap behind them just wide enough to make room for sliding shutters, and at night the homeowner simply pulled two inconspicuous handles together and turned the window into an R-12 wall.

These same techniques can be used in two additional ways to help save energy. The first is to use insulated coverings inside a solar greenhouse at night. The same clear surfaces that let sunlight into a greenhouse lose plenty of heat at night; equip your greenhouse with some sort of movable insulation to cover the glazing at night, and it becomes possible to run a solar greenhouse much more efficiently in cold weather. The other is the old medieval custom of using cloth hangings, a few inches out from the wall, to insulate an otherwise chilly space. That's what all those tapestries were doing in medieval castles; insulated wall hangings can function exactly the same way in a modern house, so long as they extend from floor to ceiling on exterior walls, and have both a reasonable amount of insulation in them and an inch or two of air space between the fabric and the wall.

None of these things are particularly difficult or expensive to make. If you have some basic facility with a sewing machine—and if you don't, getting it might be a worthwhile project—you can knock together a good set of insulated window coverings for a couple of rooms in a couple of hours, using storebought sheets and some quilt batting as your raw materials. If you know how to handle a saw, a screwdriver, and a carpenter's square, and these are also skills worth acquiring if you don't have them already, it won't take any longer to turn some lumber, hardboard, and rigid-board insulation into good sturdy insulated shutters. Examine your options, consider the ideas to be found in old and new books, and see what you can do.

Exercise for Lesson 22

For this lesson, your exercise is to figure out how much window area you have in your home. Take a measuring tape to the nearest window. Measure its height and width in inches, multiply the height by the width to get the window's area in square inches, and then divide that figure by 144 to get the size in square feet. (If you prefer the metric system, measure in centimeters, multiply as before, and then divide by 10,000 to get the figure in square meters.) Do this for every exterior window in your home, to give you a total window area. If you have single-pane windows, that entire area has an R-value of 1; if you've got double-pane windows or storm windows, it's got an R-value of 2—which is still not very much. This will give you some idea of how much heat you're losing in winter and gaining in summer through those windows, and how much you can save with some insulated window coverings.

·≈≈≈ **LESSON 23** ≈≈≈·
Conserving the Differences

The last few lessons have focused on the big picture of heat—specifically, maintaining the heat differential between the air inside your house and the air outside. The same principles also apply in more focused ways, and this lesson will focus on one of those, a cheap, simple, grubby task that most people know they ought to have done a long time ago, and a surprisingly large number of people never get around to doing: wrapping your pipes and ducts.

Here's how it works. Most American homes have a furnace and a water heater stashed away somewhere out of sight. Hot water from the water heater flows through pipes to wherever it's wanted, which may be at the far end of the house. Depending on your heating system, there are most likely either pipes taking hot water from a boiler to radiators, or ducts taking hot air from the furnace to registers, and either way they're going all through the house. That's straightforward enough.

Look at it from the point of view of thermodynamics, though, and it's a little less simple. You're trying to get a certain amount of heat from the water heater to the taps, and from the furnace to the rooms you want to keep warm in winter. To do it, you send a heated fluid, either water or air, through pipes or ducts which, for a variety of reasons, are normally made of substances that transfer heat very readily. If there's a heat differential between the fluid inside the pipes and the air outside them, in other words, you lose heat.

The standard approach to dealing with this in conventional American housing is to get the working fluid of your system hot enough so that, even after it flows through those cold metal pipes or ducts and gets to wherever it's going, you still get enough heat out the business end. That's the way Americans have learned to think about energy: the solution to every problem is to crank up the thermostat and burn more fuel. That might be a plausible approach if you've got so much of a concentrated energy source that you don't know what to do with it, and there was admittedly a time when that was more or less the case here in America, but nowadays? Hardly.

Nowadays, in a world where energy is no longer cheap and abundant, and is going to get a lot less cheap and abundant over the decades and centuries to come, we need to learn a new way to think about energy. Recognizing that energy is scarce and expensive is a good start, but it's possible to go a bit further than that, and recognize that what you need to do if you want to work with energy—especially scarce, expensive energy—is to conserve differences in energy concentration.

Your hot water pipes make a good example of this principle. The water that flows out of your water heater into your pipes is at 120°F—that's the standard setting of most water heaters these days. The air in the basement where your water heater is located is around 50°F. The second law of thermodynamics says that heat always flows from a higher concentration to a lower concentration; the rate of flow depends partly on how easily the materials in question transmit heat, and partly on the temperature differential between the two substances. In other words, when hot water flows through a cold basement, what you tend to get is lukewarm water and a basement that isn't quite so chilly. You haven't conserved the difference between the two, and the result is a chilly shower.

Insulation is one of the standard ways to conserve the difference. Wrap your hot water pipes in a good thick layer of insulation, and the heat in your hot water has a much harder time moving from the water to the air in your basement. That means, of course, that you get your hot shower. It also means that you can get the same temperature in the water coming out of the tap while using a smaller amount of energy to heat the water in the first place. That's valuable when it's a matter of decreasing the amount of fossil fuels you use, as it is for most people nowadays; it's absolutely crucial once we're talking about renewable heat sources.

As I've mentioned here repeatedly, one major difference

between the energy you get from fossil fuels and the energy you get from renewable resources is concentration. Lukewarm sunlight simply doesn't pack as much punch as burning coal. You can heat water with sunlight, but the process is never going to be as efficient as heating water with fossil fuels or electricity because the heat you get from sunlight is much more diffuse. Of course there's the additional problem that it's a bit difficult to turn up the sun twenty degrees or so to give yourself a hotter shower!

This is why "weatherize before you solarize" was one of the mantras of the 1970s-era energy conservation scene. Inefficiencies you can shrug off when you've got plenty of concentrated fossil fuel energy will cripple your attempts to make use of the diffuse heat that you can get from the sun. Fix the inefficiencies first—that is to say, conserve the differences—and you're in a much better position to begin using renewable sources. You may also save enough on your energy bills to make a solar water heater a bit more affordable.

The same logic can and should be applied in other ways. We've already talked about weatherstripping, caulking, and insulation, all of which are ways of conserving the difference between the temperature inside the house and the temperature outside. Equally, if you live in a climate with hot summers and appreciate a cool shower now and then, insulate your cold water pipes as well. If you go down into the soil more than a couple of feet you get a relatively stable, cool temperature year round; water pipes that run underground keep the water fairly close to that temperature, and if you can conserve the difference between the place your water pipe comes out of the ground and the place where it connects to your shower head, you get cold water out of the tap at, say, 60°F rather than 80°. On a sticky July day on the nether side of the Mason-Dixon line, that makes all the difference in the world.

Still, it's possible to take conserving the differences in a much broader sense, and when you do so, some very interesting possibilities unfold. Information, to return to Gregory Bateson's useful definition, is a difference that makes a difference. The capacity for energy to do work is also a function of difference—differences in temperature, pressure, electrical potential, or what have you—and so, in yet another sense, is the capacity of material substances to fill their various roles in the biochemistry of a living thing: the nutrient cycles of an ecosystem, the exchanges of goods in an economy, or the equivalent processes in any other system. In all these cases, conserving differences plays a crucial role, and there's a sense in which the degree of conservation of the various kinds of difference is a measure of the health of the whole system.

This is all the more interesting in that for some decades now, modern industrial civilization has become very poor at conserving differences. To cite only one example, American farmers used to be legendary for the facility with which they bred new varieties of any crop you care to name, specially suited to local conditions or to particular purposes. These days, by contrast, American industrial agriculture is more notable for its obsessive use of a very small number of varieties of any given crop. Plenty of factors feed into that flattening-out of differences, to be sure, but that's exactly the point: the structures that shape everyday life in contemporary America do not conserve differences—in fact, by and large, they tend to erase differences. If the conservation of difference measures the health of the whole system, our system is in trouble.

Exercise for Lesson 23

The concept of conserving differences can be taken in many directions, but the exercise for this lesson is simple—find the pipes that bring hot water to your faucets and, if you have a heating system with a central furnace, the pipes and ducts that get heat from the furnace to your home. See how much insulation they have around them. If they don't have at least half an inch, consider moving from theory to practice and wrapping them with insulation.

Your local hardware store can provide you with a variety of

materials, from foam tubes slit down one side that you can pop over your hot water lines, through fiberglass wrap that will do a good job on pressurized hot water heating system pipes, to the slightly more expensive but very effective rolls of foil-backed foam, which you can use on pipes or ducts equally well and will save you a chunk of heat. None of it's that costly, all of it can be installed by unskilled labor, and applying something of the kind will cut into your energy bills and make your home a good deal more suited to the renewable energy projects we'll be discussing in the lessons ahead.

✦ LESSON 24 ✦
Hayboxes and Sunboxes

The principle of conserving differences is central to the appropriate tech toolkit, and it can be applied in a dizzying variety of ways. One good example is a simple, resilient technology that helps solve one of the most serious problems that poor people face now and the rest of us will be facing shortly. The technology was common all over the industrial world a century ago, and you've probably never heard of it.

Let's start with the problem: cooking fuel. Most foodstuffs are safer to eat and easier to digest when they've been subjected to heat, which is why every human culture everywhere on Earth has the habit of cooking most meals. The one drawback is that the heat has to come from somewhere, and usually that requires burning some kind of fuel. Anywhere outside today's industrial world, fuel doesn't come cheap, and in most poor countries the struggle to find enough fuel to cook with is a major economic burden, not to mention a driving force behind deforestation and other ecological crises.

The obvious response, if you happen to think the way people in the modern industrial world think, is to deal with fuel shortages by finding and burning more fuel. That's exactly the thinking that got us into our current predicament, though, so it's worth looking at other options. To do that, we need to start with the thermodynamics of cooking.

Imagine, then, a saucepan on the stove cooking rice. It's a metal container with a heat source under it, and inside it are two cups of water and a cup of grass seeds—that's "rice" to you and me; the goal of the operation is to get enough heat and moisture into the grass seeds that will allow your digestive system to get at the starches, sugars, B vitamins, and other nutritious things inside the seed. So far, so good, but this is where a familiarity with the laws of thermodynamics comes in handy, because there's a prodigious waste of energy going on.

Trace the energy flow and you can watch the waste happen. The energy at the heat source is highly concentrated; it flows, with some losses, into the metal saucepan; some of it flows through the pan to the water and rice, where it does the job of cooking, but a great deal of the heat gets into the sides and lid of the pan, and radiates away into the air, so that a great deal of the energy in your cooking fuel is being used to warm the surrounding air. This is all the more wasteful in that your rice doesn't need a huge amount of heat once the water's been brought to a boil; a very gentle simmer is more than enough, but to produce that gentle simmer a lot of fuel gets burnt and a lot of heat wasted.

Here's an experiment for you to try. Get a cork mat larger than the bottom of the saucepan you use to cook rice, and a tea cozy. What's a tea cozy? An insulated cover for a teapot, designed to keep the tea in the pot good and hot while you work your way down from the first relatively pallid cups off the top to the good stuff, roughly the color and consistency of road tar, down at the very bottom. The kind of tea cozy you want has a slit in one side for the handle of the teapot, and one opposite it for the spout, and it needs to be large enough to pop over the saucepan with the saucepan's handle sticking out through one of the slits; the more insulation it has, the better.

Got it? Okay, get your pot of rice started; when the water has reached a good, fierce boil and you've put the rice in, cover the saucepan tightly, take it off the heat, put it on the cork mat and pop the tea cozy over it. Leave it for a little longer than you would normally keep it on the stove, and then serve; if you've followed the instructions, you should have perfectly cooked rice with a fraction of the fuel consumption you'd otherwise have had.

If you've done the experiment, you've just learned the principle behind the fireless cooker. In America, they were often called "hayboxes," because that's what the old-fashioned version was: a wooden box stuffed full of hay in such a way that there was a space for a pot in the middle, and a pillow of cotton ticking, stuffed with more hay, that went over the top. A hundred years ago, you could get elegant models from department stores that had porcelain-coated steel cases, rock wool insulation, and easy-to-clean metal liners with pots sized to fit; the best models had soapstone disks you could stick in the oven during the day's baking, then drop into the fireless cooker, put a pot of soup or stew on top, and have it piping hot for dinner six hours later.

I've never seen an old-fashioned fireless cooker. Here in America, most of them were turned in during the big scrap metal drives in the Second World War, though they were still in use in some corners of Europe in the 1940s and 1950s. Still, the technology is simple enough that even the least capable home craftsperson can put one together in an hour or two. My spouse and I have two of them, a portable version in a wooden box and a rather less portable version built into a piece of furniture; both of them were built using a slightly improved version of the basic haybox design, with polyester quilt batting for the insulation and cotton ticking covering the batting to keep it clean. Both of them work extremely well; the technology could probably be improved by skillful engineering, but they were cheap to make, cost nothing to use or maintain, and allow us to cook with a small fraction of the fuel that has to be used for stovetop cooking.

Those of my readers who have been paying attention may be wondering by this point if there's some way to get diffuse heat from sunlight to power a haybox. There is, and the result is called a solar box cooker. This is one of the triumphs of the appropriate tech movement: an insulated box, equipped with a window through which sunlight falls, which is entirely capable of cooking your dinner on a sunny afternoon. You can find plans for solar box cookers online or in a dozen easily accessible books, and you can make one out of cardboard, newspaper, aluminum foil, glue and a piece of discarded window glass. Placed in direct sunlight, it will easily get up to oven temperatures and cook your meals for free. It can also be used to purify tainted water, sterilize bandages, or do anything else that 300° to 400°F of steady heat will do for you.

The solar box cooker also has the not inconsiderable advantage of teaching three of the basic rules of working with solar energy in a way that most people find easy and intuitive to grasp. Rule #1 is the greenhouse effect: energy from sunlight that passes through glass and is absorbed by something inside the glass tends to get trapped there because glass is transparent to visible light but opaque to the infrared wavelengths that radiate out from warm objects. Rule #2 is the thermal mass effect: some materials absorb heat better than others, and if you put something with a high capacity for heat absorption in the presence of a heat source—say, a pot of beef stew in direct sunlight—it will soak up heat that can then be put to work. Rule #3 is the insulation effect: some materials resist the flow of heat better than others, and if you surround your thermal mass with a bunch of insulation, the heat absorbed by the thermal mass will stick around longer and do more work. Keep these three rules in mind, and most of what we'll be covering in the lessons ahead will be much easier to follow.

There are several standard designs of solar box cooker. The

simplest looks exactly like what the term suggests, a square or rectangular haybox with a pane of glass on top. A hinged lid covers the glass when the box cooker isn't in use; it has tinfoil on the underside, so that when you're ready to use the box, the lid can be propped up at an angle to reflect more sunlight into the box. Yes, you can make one out of cardboard and newspaper in about an hour, and yes, you can then set it out in the sun and cook your dinner with it.

The more complex designs put the glass pane at an angle, to increase the amount of sunlight that gets in, and they have more reflecting surfaces for the same reason. The fanciest look like metal flowers or props from 1970s science fiction movies. These, you don't want to make out of cardboard because they generally get hot enough inside to set cardboard on fire. I've seen some very impressive solar box cookers on the fancy end of things, with mirrors that would do justice to a telescope and elegant arrangements to track the sun, I'm by no means certain that all the additional complexity provides enough improvement in performance to make the effort worthwhile, but the experiment is worth making.

Two other developments of the same basic approach are also worth mentioning. The first was invented by one of the patron saints of green wizardry, the redoubtable 19th-century French solar energy pioneer Augustin Mouchot. Tasked by the French government with coming up with solar technologies for the French colony in Algeria—which had a surplus of sun and, at least in Mouchot's time, a shortage of most other energy sources—Mouchot invented, among other things, a solar cooker for units of the French Foreign Legion stationed there.[17] The device was simplicity itself—a cone of metal, reflective on the inside; a cylindrical steel cooking chamber that went up the centerline of the cone; and a tripod stand. Mouchot's solar cooker was collapsible, weighed less than forty pounds, and cooked a pot roast to a nice medium rare in under half an hour. It remained standard issue for French troops in North Africa for decades. I have no idea if any examples survive, but it's

19th century technology, so an enterprising metalworker ought to be able to knock one together fairly easily.

The other device I have in mind comes from the other end of the solar cooking spectrum, but it's as elegant as the Mouchot cooker and even more portable. The Umbroiler was invented by American solar pioneer George Löf and marketed for a while in the 1960s and 1970s.[18] It's exactly what the name suggests: a sturdy umbrella frame with silver metalized cloth in place of the usual fabric, a grill in place of the handle, and a tripod on what's normally the top of the umbrella and is the underside of the Umbroiler. It folds up like any other umbrella, but when you open it and point it toward the sun, you can cook anything from eggs to coffee on the grill. The original version was too expensive to be commercially viable, due mostly to the high cost of metalized fabric back in the 1960s; these days, that has changed, and since the patents have long since expired, a revival of the Umbroiler could make somebody with a sewing machine and basic metalworking skills quite a viable small business.

By now, I suspect, some of my more skeptical readers will doubtless be jumping up and down, eager to point out that solar cookers only work when the sun's shining. This is, of course, true, but it's also beside the point. Nothing in the appropriate tech toolkit is suited to every context—that's one of the implications of that word "appropriate," after all—and nothing ever again in human history will provide our species with the kind of instant, context-free torrent of energy we now get from fossil fuels. Once those are gone, the entire approach to technology that's built on the assumption of abundant, highly concentrated, highly portable energy supplies goes away forever, and the approaches that will replace it are going to be less convenient, less portable, and less capable of ignoring the rest of the cosmos than what we're used to.

What that means in practical terms is that a well-equipped kitchen in the deindustrial future will likely have a solar box cooker, which can be used on sunny days, and a small, efficient stove and fireless cooker, which can be used on cloudy days. It really isn't that complicated, once you grasp the crucial point that a technology that relies on diffuse renewable energy sources doesn't work the same way as a technology that relies on concentrated fossil fuels. Each form of diffuse renewable energy is well suited for some purposes and poorly suited for others; an effective application of renewables requires a locally appropriate (there's that word again) mix of technologies to do what needs to be done to make human life viable and comfortable.

Exercise for Lesson 24

As already mentioned, there are scores of books and web pages that explain how you can build a solar box cooker out of cheap or free materials. Your exercise for this lesson is to find a set of plans that suits the resources and skills you have, build a solar box cooker using those plans, and use it at least once to cook something. This isn't just an introduction to an energy-efficient method of cooking food, though it is that; more important is the practical education you'll get in learning how diffuse heat from the sun can be concentrated and put to work.

LESSON 25
Solar Hot Water

You may already be wondering if the same principles that a solar box cooker uses to cook food can also be used to provide heat for other useful purposes, such as heating the water for a hot bath at the end of a hard day's work. Yes, that can be done; it has been done, and it can still be done with remarkably little trouble and expense right now. The fact that it's being done only by a tiny minority of people in North America today does not speak well for our collective intelligence.

Solar water heating, in fact, is the most thoroughly developed renewable energy technology we have. It was perfected more than a century ago, about the same time that the internal combustion engine was being turned from a gimmick to the indispensible powerplant of the age of cheap energy. In the early 20th century, solar water heaters were the standard option in American housing straight across the Sun Belt, and they were picking up market share rapidly as a water preheater north into Canada. Only the brief heyday of cheap fossil fuel energy squeezed them out of the market. It's high time we put them back to work.

There are three basic types of solar water heater: batch, passive thermosiphon, and closed-loop active. Batch heaters are the simplest and most robust; they can be made and installed by an ordinarily competent handyperson for less than $1000 each. You already know most of what you need to know about batch heaters, because they can be described as a solar box cooker with a big black tank inside. More precisely, they consist of a tank, painted black, in a box with glass walls facing the sun and plenty of insulation everywhere else. The simple version is operated by hand: in the morning, as long as the weather is above freezing, you fill the tank by turning a tap; later in the day—the interval varies depending on location, size of tank, and intensity of sunlight—you have a tank of hot water that you can use in all the usual ways.

You can feed a batch heater into your regular hot water system, but there are more effective ways to use solar energy to heat water if this is your plan. For the time being, a batch heater is the best option for deer camps, backwoods cabins, and other situations where having cheap hot water once a day is all you need. A century or two from now, since batch heaters are the simplest and cheapest solar water heating system, they're probably the one that will be standard when the end of the age of fossil fuels has broken our descendants of the bad habit of thinking that they have a right to expect energy when, where, and how they want it. Afternoon laundry and evening baths may well be standard in that age, and it's also possible that people will also pick up the trick of running water through pipes in the back of a woodstove when it's fired up, and use the two systems jointly to keep hot water ready on tap.

Still, for the time being, if you want hot water throughout the day, you have options. If you live in a part of the world that doesn't get freezing temperatures in the winter, a passive thermosiphoning system is usually your best bet. This has a set of panels through which water flows, and a well-insulated tank located above the panels, so that hot water rises from the panels into the tank, and cooling water cycles from the tank back down to the panels. The tank connects to your hot water system—in most cases nowadays, that means that it feeds preheated water into a conventional water heater. Most of the year, if your location gets enough sunlight and your hot water system and water heater are adequately insulated, the solar panels will do all the work of heating water for you; in the cold part of the year, you'll need additional heating, but even so, the water that reaches your heater comes in already warm, so the amount of energy needed to heat it the rest of the way to comfortable bath and dishwater temperature is a fraction of what you would need to heat it all the way from cold.

In areas that freeze in the winter, the standard approach is a closed-loop active system that uses something other than plain water to take heat from the panels and circulate it into the tank. Various antifreeze solutions are standard; they go from the panel to a heat exchanger in a well-insulated tank, and a small electric pump keeps the fluid moving. Once again, heat goes from the panels to the insulated tank, hot water from the insulated tank goes into the regular water heater, and keeps it from having to work hard or, under good conditions, at all. On average, a thermosiphon or closed-loop system will provide you with 70% of your hot water free of charge, though that figure varies significantly by location; in Sun Belt locations with plenty of clear skies, it tends close to 100%. Since water heating accounts for around 15% of the average household energy bill, a solar water heater in an average location will provide around 10% of your total home energy use forever.

Batch heaters are fairly easy to build, but thermosiphon and closed-loop systems are not, and plenty of people back in the 1970s found out the hard way that good intentions are not an adequate substitute for the right tools, materials, and experience. For most people nowadays, buying a solar hot water system is the better option. The cost for systems of this kind varies widely depending on the details of the system, the orientation of your house toward the sun, and the ease with which pipes can get from the solar system to your ordinary hot water system. As previously mentioned, you can get a batch system in place for $1000; $4000 is a good ballpark figure as of this writing for a passive thermosiphon system, and plan on about $8000 for a closed-loop active system. It's a noticeable upfront cost, to be sure, but it's worth remembering that there are no additional fuel costs for as long as the sun is well stocked with hydrogen.

There are probably other ways to heat water using the sun that haven't been invented yet, but the three methods that I've mentioned here have a major advantage: they are mature technologies, with strengths and drawbacks that have been tested thoroughly in practice. When you install a solar water heating system, you know exactly what you're getting into, and, though it can sometimes be necessary to correct for the enthusiasm of salespeople, most of the firms that install solar hot water systems have been around for decades and have learned, as most small businesses learn, that satisfied customers are the advertising that matters. You can also look up the performance of the available models in independent sources, go fishing for complaints on the internet, and do everything else you would normally do when assessing a piece of technology for your home.

Like hayboxes and solar box cookers, solar water heating systems are simply one more way of conserving thermal differences or, to make the same point in a somewhat different manner, one more way to tap into freely available, renewable sources of diffuse heat to do things that most people in the industrial world do by burning up irreplaceable fossil fuels. Still, it's important to remember that hot water is not merely a luxury; the ability to wash dishes, clothing, and bodies in hot water on a regular basis has done a great deal to get rid of many health problems that were common among our ancestors, and being able to do the same thing in a post-petroleum future will do the same thing for our descendants—if the technology gets to them. The best way to get it to them is to revive it and use it now.

Exercise for Lesson 25

One consequence of industrial society's fixation with using highly concentrated energy sources is that very few people have a clear sense of how much diffuse heat can be collected from sunlight. Your exercise for this lesson is to learn about that by direct experience, and, in the process, get a basic introduction to how a batch solar water heater works.

To do this, you'll need a glass bottle, as large as you can conveniently find, and a kitchen thermometer—the kind that will measure temperatures between room temperature and the boiling point of water, and sized so that you can stick it in the mouth of the bottle. Paint the outside of the bottle flat black and fill it with water; leave a couple of inches of air at the top, measure the temperature of the water with your thermometer, and then fasten the lid on tightly and put the bottle in the solar box cooker you made for the previous lesson's exercise. Leave it in full sunlight for an hour; then take it out using hot mitts—the glass will be hot enough to give you a nasty burn—open it carefully, keeping your bare skin and eyes out of the way of any steam that may spurt out, and measure the temperature of the water. Can you use this same method to provide yourself with modest amounts of hot water in an emergency? If the sun's shining, you bet.

⟞⟝ LESSON 26 ⟝⟞
Passive Solar Heat

The same techniques that have been used in previous lessons to heat food and water can also be used to heat air, and the 1970s appropriate-tech movement made use of that fact to come up with a variety of clever ways to heat homes with the sun. Two broad approaches shaped the schemes devised in those days. The first was passive solar heating, the second was active solar heating; the difference between the two is precisely the same as the difference between passive and active water heating—does the system get heat where it needs to go without additional energy input, or is a pump needed to move heat from place to place?

Passive solar heating was far and away the most popular approach, and some very impressive systems were devised to use it. In most areas, given a decent southern exposure, a house designed for passive solar heating, and adequately insulated and weatherized to make best use of it, requires little or no heating other than what the sun provides. The one drawback, and it's a significant one, is that the house has to be designed and built from the ground up with passive solar heating in mind. Those of my readers who expect to have the resources to build a passive solar house, or have one built for them, should certainly look into the technology; the rest of us will be living in existing construction, and the possibilities here are more limited.

The most important limit, of course, is that you can't do passive solar at all unless a good part of the south or southeast face of your house receives direct sunlight during at least a significant fraction of each winter, spring, and autumn day. Some houses have that option; many others don't, and if you don't, you need to do something else. If you have adequate sunlight, on the other hand, you have at least three options available, and they can be used alone or together.

The first is a thermosiphon air panel (TAP). If you remember how a passive thermosiphon solar water heater works, you already know most of what you need to know here. A TAP is a wide, flat box with glass on the front, insulation on the sides and back, and a sheet of metal running parallel to the glass, with a couple of inches of air space between metal and glass. Air comes in at the bottom, flows over the metal, and goes out the top into the space that needs to be heated. Position the panel in the sun, and the metal very quickly gets hot; the air passing over the panel picks up the heat, and you very quickly have cold air being sucked into the pipe that leads to the bottom, and hot air being blown out the pipe that leads out of the top.

The TAP is one of the cheapest solar technologies you can make—it costs about as much as a good solar oven—and it's also portable; I've seen TAPs in rental properties rigged to lean against a southeast wall outside and feed air inside through ducts sized to fit a window open at top and bottom. Furthermore, it produces heat fast: if you live someplace where winters are cold but sunny, and you can place the panel so that it catches rays as soon as the sun comes up, you can have hot air warming your house within a half hour or so of dawn. The downside is that the heat goes away as soon as the sun does, and at night, the thermosiphon effect can work in reverse—hot air gets sucked in the top, flows over the chilly metal, and emerges as an icy breeze at floor level. Thus a TAP needs valves to cut off the air flow when the sun goes away. If you've got a well-insulated and thoroughly weatherstripped house, the heat from a couple of well-placed panels can keep you comfortable well into the night, but the technology does have its limits for round-the-clock heating.

To balance the quick but unsteady heat of a TAP system, you need another system that soaks up heat whenever the sun is out and distributes it to the house in a steadier manner throughout the day and night. The key to this effect is thermal mass. Some substances are good at soaking up heat; when it's hot, they absorb it, and when it gets cold, they radiate it. Old-fashioned fireplaces used to include plenty of brick or stone precisely because these have plenty of heat storage capacity and radiate heat via infrared rays long after the fire has been banked down for the night. In the same way, most passive solar systems use plenty of thermal mass to soak up the sun's heat in the daytime, and radiate it all night.

There are several different gimmicks for retrofitting a house to use thermal mass. One of the standard methods, back in the day, was the trombe wall. What's a trombe wall? Basically, it's a wall-sized TAP with thermal mass rather than a metal sheet inside the glass. One very effective, though rather ugly, way of building a trombe wall was to take black 55-gallon drums full of water and stack them in a sturdy frame so that their ends faced the sunlight; glass went over the sunward surface, a few inches from the ends of the drums, and the wall on the other side was pierced by vents at top and bottom, which could be opened and closed. Some kind of insulation to cover the glass on a cold night or cloudy days was a common addition that improved the efficiency of the system quite a bit. Water is among the very best thermal masses, but brick, stone, or concrete will also do a good job; the less unsightly trombe walls tended to use these instead of barrels of water.

The next step up from the trombe wall, and one of the most widely used and thoroughly tested of the passive solar retrofit technologies, is the attached solar greenhouse. You build this onto the south or southeast face of your house, sealing it up tight so that air doesn't leak in or out, and put a trombe wall between the greenhouse and the rest of the house; the floor of the greenhouse may also be made of heat-absorbing brick, stone, or concrete, to add to the effect. Sunlight streaming in through the glazing warms the air and the trombe wall inside, and heat then radiates from the thermal mass to the rest of the house, regulated by vents that can be opened or closed; the greenhouse should also be vented to the outside on hot days. In addition to a significant heat gain, of course, the greenhouse also allows you to keep fresh vegetables in the diet from early spring into late fall, and right through winter in climates that aren't too arctic.

Those are the most successful passive methods. Quite a few experiments were made with active solar space heating—that is, systems that collect heat from the sun and then pump it somewhere else. It can be done, but because of the diffuse nature of solar heat, the efficiencies are low, and you can very easily end up using (and losing) more energy in the process than you gain by it. That's been a persistent problem all along with attempts to run complex systems on the diffuse and intermittent energy flows that can be gotten from renewable sources. Too many people, faced with that reality, either give up on renewables altogether, or waste their time and resources trying to find some gimmick that will allow a diffuse and intermittent energy source to do the same things as a concentrated and instantly available one.

Given that renewables are the only energy supply we can count on for the long term, the first choice is not very helpful. Given that the laws of nature are under no compulsion to provide our species with the kind of energy supplies that the fraction of humanity currently living in industrial societies seem to think they are entitled to get, the second choice is not much better. The viable alternative, of course, is to recognize that renewable energy sources can't simply be shoved into existing roles as replacements for oil, coal, and natural gas; they require different ways of thinking about energy, and they imply an entirely different kind of energy technology.

The thermosiphoning air panels, trombe walls, and attached solar greenhouses that emerged as the best products of a decade of lively experimentation are baby steps in the direction of the energy systems of the far future. Still, just as baby steps are precisely what's most appropriate when a baby starts learning to walk, these simple, flexible, and inexpensive approaches are good ways to make a start on the task of learning how to live comfortably on the diffuse energy flows nature provides. It's also important to remember that all these things can be put to use by individuals, families, and local community groups with readily available resources, very much including salvage—old windows, for example, make excellent glazing for all three of the systems just discussed.

(Exercise for Lesson 26)

At intervals during the next convenient sunny day, go outside and see what parts of your house receive direct sunlight. Is there any place where a TAP, a trombe wall, or an attached solar greenhouse might be convenient to install? Research those technologies, using the resources for this section and any other information source you can find, and consider installing one or more of them, if you can, to provide at least part of your home's heat from the sun.

✦✦✦ LESSON 27 ✦✦✦
Solar Electricity

Talk about solar energy these days, and most people who listen at all will likely assume that you're talking about some method of turning sunlight into electricity to feed the power grids that keep today's extravagant lifestyles powered. We've already discussed the reasons why today's regional and continental grids are likely to go the way of the dinosaur and the dodo, and, although it's already been stated, this is worth repeating: in most cases, the best thing to do with solar energy is to let it convert itself into diffuse heat and use the heat.

Still, electricity is a useful thing to have, and there are some applications—in particular, communications and other electronics technologies—where nothing else will do. Thus it's reasonable to consider the potentials for producing modest amounts of electricity for those specialized applications. Making electricity from sunlight is certainly one option, but it's a complex option, because most of the options for home-generated electricity available right now fail the sustainability test in one way or another.

Photovoltaic (PV) power might as well be the poster child for this effect. PV chips are made by a variant of the same process that produces computer chips, and they face the same problems with complexity, which will become more and more significant as the economic and technological basis for fabrication plants and world-wide supply chains comes unglued. Though silicon, the raw material in most PV chips, is one of the most abundant elements on the planet, many of the other substances used in manufacturing solar panel systems are much scarcer, and there are also major issues with toxic wastes and other pollutants, so there are significant ecosystem limits to the technology as well. All things considered, it's probably a safe bet that within fifty years or so, PV cells will no longer be manufactured. Does this mean that PV panels should be off the list for green wizards today?

That depends on what your PV panels are intended to do, for there are two sides to the challenge that green wizardry is intended to meet. The first and most obvious task before us is to begin the process of creating and deploying prototype versions of sustainable lifestyles, homes, and communities, on a scale small and local enough that the inevitable mistakes and mischances can be managed. The second, which is too often neglected in discussions of the subject, is to meet the needs and reasonable wants of the people who are doing all this creating and deploying. Both these tasks, furthermore, must be done in an age of economic technological unraveling when relying on the continued functioning of today's massive and centralized systems is a sucker's bet.

The fact that those panels won't be available fifty years from now doesn't make them useless today, and someone whose main efforts are directed toward organic gardening, say, or some other dimension of the green wizard project, could do a lot worse than to cut her electricity use down to size and then provide the current she needs from a bank of solar panels and a stack of batteries. For that matter, even someone who's hard at work in the basement lab assembling some other source of home electricity might choose to retool his lifestyle in the meantime to work off a hundred watts or so of 12-volt power and put up a few PV panels to provide that power while tinkering with the generator and getting it through the teething pains every experimental project gets to enjoy.

That is to say, PV panels can be used as a bridge. Unlike the natural gas being pumped out of the ground so frantically by fracking operations just now, it's a bridge that leads somewhere—or, more precisely, it has the potential to lead somewhere, though it can also be used in less helpful ways. The sort of grid-tied PV panel system that's designed to feed 110 volts of alternating current into the grid and can't be used at all when the grid goes down—yes, there are plenty of PV installations like that these days—is a bridge to nowhere; it's designed to prop up a way of life with no future, or more precisely, to go through the motions of propping up that way of life, while serving primarily as a status symbol.

The land on the other side of the bridge, to extend the metaphor a bit further, will inevitably be a place where the inhabitants use a lot less electricity than people in the industrial world do today. Just as you need to weatherize before you solarize, to repeat the motto from the seventies, you thus need to make very serious cuts in your electricity use before you can realistically turn to renewable sources to meet the modest power needs that remain. Once those cuts are made, the range of possible options to provide the little electricity you actually need is considerably wider than it looks at first glance.

One example that gets very little attention these days is solar thermoelectric power. Imagine a zigzag of metal in which, so to speak, all the zigs are all of one kind of metal (say, copper), all the zags are another (say, zinc), and the two metals join at the angles. If you heat the angles on one side of the zigzag and cool the angles on the other side, electric current starts flowing through the zigzag, and if you solder wires to the two ends and connect them to something that uses electricity, you're good to go. On a small scale, it's a surprisingly robust effect; back in the 1940s and 1950s, Russia used to manufacture sturdy little thermoelectric generators that put the heat from a kerosene lamp on one side of the zigzag and the Siberian climate on the other. Those proved quite adequate to power the tube-based radio receivers standard at the time, which weren't exactly abstemious in their power needs.

Back in the day, a certain amount of tinkering was done on solar thermoelectric power, and it turned out to be very useful for satellites because the heat differential between a lump of hot radio-active metal and the chill of interplanetary space produces a nice, steady current suitable for deep space probes. Its possibilities on an industrial scale never amounted to much, though, as it proved to be difficult to scale up to any significant degree, and, of course, as long as we can count on a steady supply of cheap abundant fossil fuels, solar thermoelectric power is a non-starter.

Look past the tyranny of temporary conditions, though, and the possibilities are fascinating. To say that a solar thermoelectric generator is a simple device understates the case considerably. Benjamin Franklin could have knocked one together in a spare afternoon while waiting for the next thunderstorm to blow in; for that matter, it would not have posed a significant challenge to a skilled craftsperson in ancient Egypt. All you need is the ability to work nonferrous metals and the very basic geometry skills needed to shape a parabolic dish reflector. Strictly speaking, the efficiency of heat-to-electricity conversion isn't that high, but given a more useful definition of efficiency—for example, labor and resources input to electricity output—it leaves many other options in the dust, and its sustainability is hard to match; we're talking about a technology that could power radio communication and the like for as long as our species endures.

Down the road, solar thermoelectric generators are likely to become one of the standard ways that households and small busi-nesses provide themselves with a modest supply of electricity, while PV panels will be an exotic legacy from the industrial past where they've survived at all. There's a fair amount of road to be covered between now and then, however, and during much of that time, those solar thermoelectric generators will be making the journey that runs from handbuilt prototypes in the backyards of basement-workshop inventors, through balky first-generation models of many different designs turned out by green entrepreneurs on shoestring budgets, to the shaking-out process from which the standard, sturdy, widely available models of the future will finally emerge.

During that time, those of my readers who don't happen to have a talent for nonferrous metallurgy and electrical engineering may find PV panels a useful investment. There are quite a few potential bridges that lead somewhere, just as there are other technologies that aren't bridges at all, but fully sustainable options that will still be running long after the last PV cell stops working. In a world where the industrial nations didn't take a thirty-year break from reality, it probably wouldn't be necessary to use the bridges at all.

Still, that's not the world we live in. The world we live in is one in which a small minority of people are belatedly waking up to the predicament into which the misguided choices of recent decades have backed us, while most others are squeezing their eyes shut and covering their ears with their hands in a desperate attempt to keep from noticing the mess we're in. In that kind of world, saving much of anything at all is going to involve quite a bit of last-minute scrambling and a fair number of temporary expedients and jerry-rigged makeshifts. One feature that will likely be common to a great many of those latter is the use of resources extracted in one way or another from the disintegrating mass of our current industrial system.

Exercise for Lesson 27

Go back over the list of electrical appliances you made for Lesson 19's exercise, and this time, rate them in terms of importance. Next to each of them, mark down an N if it's a necessity, a C if it's a convenience, an L if it's a luxury, and a W if it's simply a waste of current. When you're done, find out how many watts each of the things on your N list uses—that figure can be found somewhere on all electrical appliances—and estimate from that the maximum number of watts you need in the course of the day to keep your necessary appliances running, remembering that some appliances need to run all the time while others get turned on and off. Then, using online solar calculators or other resources, work out how large of a solar PV system you would need to provide that much electricity, and approximately how much it would cost you to have such a system installed. If the figure seems too high, look over your list again and see if some of the "necessities" don't deserve that name.

⋰⋰⋰— LESSON 28 —⋱⋱⋱
Wind Power

The logic applied in the previous lesson to photovoltaic solar power can be applied more generally. There are certain technologies that are either dependent on electricity, or are easiest to provide using electricity, that contribute mightily to human welfare; long-range radio communication is an example of the first kind, while refrigeration for food storage is an example of the second. If these technologies can get through the end of the industrial age in a sustainable form, they will contribute to human welfare as far into the future as you care to look. Renewable energy sources that provide a modest amount of electricity on a local scale can keep a good many of these technologies going, and if enough people either learn how to build and maintain renewable systems on that scale, on the one hand, or learn how to build and maintain the technologies themselves on the same modest and local scale, on the other, our civilization may accomplish the surprisingly rare feat of adding something worthwhile to the long-term toolkit of our species.

The modest amount and the local scale are vital to any such project. Right now, anyone with a fairly good set of hand tools and a good general knowledge of electricity, carpentry, and metalworking can build a wind turbine for a few hundred dollars. I can say this with some confidence because I helped do exactly that, for a good deal less, while attending college in the early 1980s. The turbine itself was basically a two-blade propeller cut, shaped, and sanded from a chunk of spruce; the conversion of rotary motion to electricity was done by an alternator salvaged from an old truck; the tail that kept it facing into the wind, the safety shutoff that swung it out of the path of the wind when the wind velocity got too high, and the tricky doodad that allowed it to turn freely while still getting electricity down to the batteries in the little shed at the base, were all fabricated out of scrap parts and sheet metal. We borrowed a disused power pole to put the turbine up where the wind blew freely, but if that hadn't been there, a tower could easily have been put together out of readily available hardware and bolted onto a hand-poured concrete foundation.

The design we used wasn't original, not by a long shot. Half a dozen old appropriate-tech books from the Seventies have the same design or its kissing cousin; it's one of a half dozen or so standard designs that came out of the ferment of those years. The most important difference was between the horizontal axis and vertical axis models. A horizontal axis wind turbine is the kind most people think of, with blades like a propeller facing into the wind and a tail or some other gimmick to pivot it around in the right direction. A vertical axis wind turbine is less familiar these days, though you used to see examples all over the place back in the day; the business end looked either like one side of an eggbeater—the Darreius turbine—or an oil drum cut in half lengthwise, and the two sides staggered around the vertical shaft—the Savonius turbine. Some of the standard designs yielded high speed and low torque, which is what you want for generating electricity; some of them produced high torque and low speed, which is what you want for pumping water or most other uses of mechanical power.

All the information needed to design and build one or more of the standard models is easy to come by nowadays—literally dozens of books from the time cover the basic concepts, and it's far from hard to find detailed plans for building your own. It's also not too difficult for those who lack the basic technical skills to find small wind turbines of quite respectable quality for sale, though the price is going to be a good deal more than you'd shell out for an old truck alternator, a chunk of spruce six feet by eight inches by four inches, and the rest of the hardware we used to cobble together our turbine. Either way, if you live in an area with average winds and your home isn't surrounded by tall trees, steep hills, or skyscrapers, your odds of being able to run a respectable 12-volt system are pretty good.

Still, it will come as no surprise to my readers that very little of this wealth of practical information receives much in the way of attention nowadays. Instead, the concept of wind power has been monopolized by a recently minted industry devoted to building, servicing, and promoting giant wind turbines that provide electricity to the grid. The giant turbines have their virtues, no question; compared to most other energy production technologies, certainly, they're relatively safe and clean, and their net energy yield is a respectable 8 or 9 to 1, which beats the stuffing out of most other alternative energy sources. Still, the idea that serried ranks of giant wind turbines will enable us all to keep on using energy at today's extravagant rates runs headlong into at least two difficulties.

The first difficulty is intermittency. A wind turbine, obviously enough, produces power only when the wind is blowing, and it's a safe bet that no matter where you put turbines, the wind won't always be blowing. That wouldn't be a problem at all if Americans were used to using electricity when it happens to be available, and doing something else with their time when it's not, but that's not the way Americans do things. Just now, intermittency isn't much of a problem because modern gas-fired power plants can be cycled up and down promptly to respond to any shortage of power from the turbines. If, however, your plan is to replace the gas-fired plants (and the coal-fired ones, which can't be cycled up and down so quickly) with wind turbines, you've got a problem. You have an even bigger problem if you want to rely on solar as well as wind, since then you're dependent on two intermittent energy sources. When they both go down at the same time—as, by Murphy's law, they inevitably will—you're left with no power going into the grid at all.

The second difficulty, as discussed previously, is complexity. Those giant turbines, it bears remembering, are not made out of spare truck alternators, blocks of spruce, and other readily accessible and easily managed parts. They are triumphs of modern engineering, which means in practice that they depend on baroque supply chains, high-tech manufacturing processes, and massive investment, not to mention plenty of fossil fuels and, more generally, a society that has plenty of cheap energy to spare for projects on a gargantuan scale. Nor is a giant wind turbine sitting all by itself on a hilltop particularly useful to much of anyone; it gains its economic viability through connection to the electrical grid, which is itself an immense technostructure with its own even more sprawling supply, manufacturing, and investment requirements. If industrial society finds itself unable to maintain any one of the factors that make the grid and the giant turbines possible, then it doesn't matter how useful they might be; they won't exist.

Homescale windpower systems suffer from the intermittency issue, but then so does nearly every other option for providing electricity on that scale, and we've already discussed at some length the solution to it: get used to using electricity when it's available, or to storing up modest amounts of it in inexpensive storage batteries and using that supply sparingly. The challenge of complexity, on the other hand, is not something a homescale windpower system has to deal with at all. Even in the absence of salvageable alternators, and there are quite literally hundreds of millions of them lying unused in junkyards across the United States, a generator that will turn rotary motion into direct current is not a challenging project. I built a simple one in elementary school, for example, and although it wasn't really suited to wind turbine use—most of the structural elements were made from paperclips, with a toy horseshoe magnet to provide the field, and the amount of current it produced was just about enough to get a decent glow out of a very small light bulb—the principle can readily be scaled up.

In the kind of future we can realistically expect, in other words, homescale windpower will almost certainly be a viable technology, while giant wind turbines of the modern sort won't. Now, of course, it's a safe bet that the windpower industry as it now exists will keep on building, servicing, and promoting giant wind turbines as long as it's possible to do so, so the small chance that the giant turbines might actually be viable is covered. What isn't covered yet is the very large chance that small wind turbines of the sort that can be built and maintained in a basement workshop could provide a real benefit during the difficult decades ahead of us.

In order to respond to that range of possibilities, homescale windpower units need to find their way back into the conversation of our time and, more importantly, up above the rooftops of homes across the modern world. Professionally manufactured wind turbines of the right scale are a good start, and those green wizards in training who have the money and lack the fairly modest technical skills to build their own could do worse than to buy one. Still, there's also a role here for the homebuilt turbine. For those individuals whose willingness to get to work shaping turbine blades and bolting together towers, there might be, as things unfold, possibilities for a future career.

Promoters of giant wind turbines, and for that matter of centralized power generation schemes of all kinds, tend to talk quite a bit about economies of scale. In an expanding economy with a stable or growing resource base, that sort of talk often makes sense, though the extent to which those economies of scale are a product of direct and indirect government subsidies to transportation, financing, and large businesses generally is not something economists like to talk about. Still, in a world facing economic contraction, resource depletion, and a loss of complexity potentially capable of rendering a great deal of today's infrastructure useless or worse, the balance swings the other way. In the face of a future where small, cheap, localized approaches that are sparing in their use of resources, relying on massive, expensive, centralized, resource-intensive power plants of any kind is not an economy but a profligacy of scale—one that we very probably will not be able to afford for much longer.

(Exercise for Lesson 28)

The space needs of a homescale turbine are very simple: it needs to be thirty feet higher than any obstacle within a hundred feet in any direction. For the exercise for this lesson, go outside and see whether there's any way that such a turbine could be installed where you live. Apartment dwellers have an advantage here—it will be a source of endless puzzlement to our descendants that we didn't use those nice tall buildings as bases for wind turbines—but most American homes could readily have a 12-volt homescale wind turbine mounted onsite. One way or another, figure out where, if you decided to install a turbine, it might best be placed.

LESSON 29
Other Energy Sources

Solar energy and windpower were the two sources of energy that got most of the attention during the heyday of the appropriate tech movement, and they are the two I worked with most back in the day, but they were never the only options. One of the core principles of the movement was the recognition that every place and every situation has its own mix of potential energy sources. A number of the other common options were explored by appropriate tech wonks, and deserve discussion here.

"Micro-hydro" and "mini-hydro," for example, are potentially options of great importance in the broad picture of a post-abundance energy future. The "hydro" in each of these phrases, of course, is short for "hydroelectric." Micro-hydro is homescale hydroelectric power, usually produced by diverting a small amount of a stream or river on one's property through a small turbine to spin a generator. Back in the day, there was a certain amount of work done with simple undershot waterwheels—this is the technical term for a waterwheel in which the water flows underneath the axle, which is far and away the simplest kind to make. Most of the 1970s-era designs were made from scrap metal and hooked up to truck alternators of the same sort used in so many homebuilt wind turbines; I have no personal experience with how well these worked, but the concept is worth revisiting.

Mini-hydro is the next step up—hydroelectric power on the scale of a neighborhood or a rural town. Unlike today's mega-hydro plants, this doesn't require damming up whole river basins, devastating fish runs, and the like. A small portion of a river's flow or a small and steep stream provide the water, and the result under most circumstances is a supply of sustainably generated electricity that doesn't suffer from the intermittency of sun and wind. Of course, it depends on having the right kind of water resource close by your community, and that's a good deal more common in some areas than others; it also requires a good deal more investment up front. If you can get past those two obstacles, it's hard to think of a better option.

Small amounts of electricity can be generated in a variety of other ways—for example, by burning something and using the heat to run a thermoelectric generator of the kind discussed in Lesson 28. Still, one of the core rules of appropriate tech is that the thermodynamic costs of turning some other form of energy into electricity, and then turning the electricity back into some other form of energy such as rotary motion or heat, can be ignored only if you've got a half billion years or so of stored sunlight to burn. There are situations where those losses are worth accepting, but not that many of them. If you can leave the energy in its original form and not take it through the detour into electricity, you're usually better off.

Methane is an example. Methane production from manure on a small scale is a going concern in quite a few corners of the Third World. You need more raw material than a single human family will produce to get a worthwhile amount of gas, but small farms with livestock yield enough manure to keep a small kitchen stove fueled on this very renewable form of natural gas. (The residue makes excellent raw material for compost, since only the carbon and hydrogen are involved in methane production; the nitrogen, phosphorus, potassium, and other plant nutrients remain in the sludge at the bottom.) Since fuel for cooking is higher on the list of basic human necessities than most things you can do with modest amounts of electricity, this is probably the best use for the technology.

Flatulence jokes aside, I have no personal experience with small-scale methane production. Wood heat, on the other hand, is a technology I've worked with, and it's probably going to be a major factor in the energy mix in North America in the future. It's a simple, robust technology that works very well on the home scale—in fact, it's hard to use it on any larger scale—and many wood stoves come with what's called a waterback, which uses heat from the stove to heat domestic hot water. (Combine solar water heaters with a cooking stove equipped with a waterback, and you've got your hot water needs covered year round.) The problem here is that wood heat is a major cause of deforestation worldwide. Too much harvesting of wood has devastated ecosystems over much of the world and caused a range of nasty blowbacks affecting human as well as biotic communities.

There's at least one way around that problem, though it needs to be implemented soon and on a large scale. A very old technique called coppicing allows for intensive production of firewood off a small acreage. The trick to coppicing is that quite a few tree species, when cut down, produce several new shoots from the stump; these grow much more rapidly than the original tree, since they have their root system already in place. When the shoots get to convenient firewood size, the coppicer cuts them again, and yet another set of shoots come up to repeat the process. I've dabbled in coppicing: the vine maple of the Pacific Northwest, which grows like a weed and produces decent firewood, made that easy enough, and other regions have their own equivalents. As other fuels run short, the owner of a few acres who uses it for coppicing and sells dry wood nicely sized for wood stoves may have a steady income, or at least a perennial source of barter, for the foreseeable future.

Biofuels such as ethanol and vegetable oils are another source of heat energy that will probably see a great deal of use in the future, though here again the limits on production are not always recognized. In a world with seven billion mouths to feed and an agricultural system at least as dependent on fossil fuels as any other part of industrial civilization, diverting any large portion of farmland from growing food to producing biofuels risks a substantial political backlash. I wonder how many of the proponents of biofuels production have thought through the consequences of a future in which the hazards of driving might include being stopped at makeshift barricades and torn to pieces by an impoverished mob that is all too aware that every drop of ethanol or biodiesel in the tank represents food taken from the mouths of their children.

Biofuels are likely to play some role in the early stages of the end of the age of abundance, but thereafter, at least until the world's human population and post-petroleum agriculture have settled down into some sort of equilibrium, it's unlikely that this role will be very extensive. Later on, it's anyone's guess, but the answer will be up to the people of the twenty-fourth century, not us.

Methane, wood, and sunlight, then, will probably account for the great majority of heat energy in common use in the centuries immediately ahead of us. What about mechanical energy? The breakthrough that launched the Industrial Revolution was the discovery that heat from burning coal could be turned into mechanical energy by way of a steam engine, and much of what sets our civilization apart from other civilizations in history is precisely the ability to put almost unimaginable amounts of mechanical energy to work. If a car with a 100-horsepower engine literally had to be pulled by a hundred horses, for example, and each of those horses required the care and feeding that horses do, the number of such cars on the roads would be a very small fraction of the present total.

There are good reasons, some historical and some pragmatic, to think that the major source of mechanical energy in the post-abundance future will be what it was in the pre-abundance past, that is, human and animal muscle, amplified by a variety of clever tools. If anything, some of the more ingenious inventions of the last few centuries make muscle power even more useful now than it was before the first steam engine hissed and groaned its way into a new age of the world. The extraordinary efficiency with which a bicycle converts muscular effort into movement is a case in point. The relatively simple metallurgy and engineering needed to build a bicycle is likely to survive into the far future, or to be reinvented after some more or less brief interval. The sheer value of a technology that can move people and supplies a hundred miles a day on decent roads will hardly be lost on our descendants. It's far from unlikely, for example, that wars will be won in the post-petroleum era by those nations that have the common sense to equip their infantry with bicycle transport.

More generally, the invention of really effective gears may turn out to be one of the nineteenth century's great contributions to the future. The Roman world had complex machines using cogs and gears, but the designs used at that time did a poor job of transmitting power; gearing systems that originally evolved in the late Middle Ages for clockworks underwent dramatic changes once steam power created the need to transfer mechanical motion as efficiently as possible—from place to place and from one direction to another. Once invented, effective gears found their way back down the technological pyramid to the realm of hand tools; anyone who has ever compared beating egg whites with a spoon to doing so with a hand-cranked beater will have a very clear idea of the difference in effort that such simple mechanical devices make possible.

That difference may not seem like much in comparison to the gargantuan achievements of current fossil fuel-powered technology, or the even more grandiose fantasies served up by those who insist that the end of the age of petroleum must, by some kind of technological equivalent of manifest destiny, usher in the beginning of the age of some even more titanic energy resource. Still, it's past time to shelve the fantasies of limitless energy and the hubris that goes with them. We must start paying attention to the tools, technologies, and modest but real energy sources that can actually have a positive impact on human existence in an age when only natural phenomena have gigawatts at their disposal any more.

Exercise for Lesson 29

For this lesson, your exercise is to consider the potential homescale and community-scale energy sources in the place where you live. Online sources and books on renewable energy can tell you roughly how much sun and wind you can expect to get in an average year, and that's a good start. Beyond that, does your home or your community have a nearby stream or river that could be used for micro-hydro or mini-hydro? Is livestock raised in the region? What kind of trees grow readily in your area, and are any of them suitable for coppicing? What kinds of draft animals did people in your area use before the introduction of tractors and automobiles, and does anyone nearby still know how to use them? Try to get as complete a picture as you can of the energy resources available where you live.

··· LESSON 30 ···
Transportation

There is a major difference between the kinds of energy resources and energy use I've been discussing and the kinds that get most of the attention in the mainstream press—when it mentions energy issues at all. No, I'm not talking about the distinction between centralized grid-based electricity and homescale production of diffuse heat, though that's an important issue. What I have in mind is the way I've avoided talking about the energy resource that gets most of the attention from the public and the political sphere alike—and that is, of course, the gasoline that goes into the tanks of five hundred million or so North American automobiles.

Some people in the old appropriate-tech movement dabbled in alternatives to gasoline, mostly by way of inefficient electric cars and very efficient bicycles. Nowadays, a dozen large and small automakers are bringing out battery-powered cars of various kinds, ranging from compacts like the Ford Focus and Nissan Leaf to exotic items like the Aptera and the GEM. They range in price from high to astronomical, and all of them have their share of drawbacks, mostly in terms of range and reliability, but a significant number of people on the green end of things are hailing the appearance of these vehicles as a great step forward. As things stand, that's a bit of an oversimplification. Most all the electricity these vehicles use will be generated by burning coal and natural gas, and the easy insistence that the grid can easily be switched over to solar and wind power won't stand up to close analysis. These points have been discussed to some extent in the alternative energy scene, but there are other points that deserve at least as much attention.

First of all, the best way to reduce your ecological footprint isn't to replace a gasoline-fueled car with an electric car. It's to replace it with a bicycle, a public transit ticket, or a good pair of shoes. Now, of course, the built geography of much of rural and suburban North America makes it a little challenging to do without a car, but close to a hundred million people in the United States live in places where a car is a luxury most or all of the time, and a significant portion of the others choose to live in places where that's not the case. Still, let's set aside for the moment the fact that the one energy-related bumper sticker that might actually make a difference would go on the back of a bicycle, and would say "MY OTHER CAR IS A PAIR OF SHOES." For those Americans who actually do need a car, how about the new electric vehicles? Will they really decrease your carbon footprint and your fossil fuel use, as so much current verbiage claims?

The answer is no. First of all, as already mentioned, the vast majority of electricity in America and elsewhere comes from coal and natural gas. Choosing an electric car simply means that the carbon dioxide you generate comes out of a smokestack rather than a tailpipe. The internal combustion engine is an inefficient way of turning fuel into motion—about three quarters of the energy in a gallon of gas turns directly into low-grade heat and gets dumped into the atmosphere via the radiator, leaving only a quarter of the total to keep you rolling down the road. The processes of turning fossil fuel into heat and heat into electricity, storing the electricity in a battery and extracting it again, and then turning the electricity into motion is less efficient still, so you're getting less of the original fossil fuel energy turned into distance traveled than you would in an ordinary car. This means that you'd be burning more fossil fuel to power your electric car even if the power plant was burning petroleum, and since coal and natural gas contain much less energy per unit of volume than petroleum distillates, your electric car is burning quite a bit more fossil fuel, and dumping quite a bit more carbon in the atmosphere, than gasoline-powered cars do.

This isn't something you'll see discussed very often in e-car websites or sales flyers. It's even less likely that you'll find any mention there of the second factor that needs to be discussed, which is the energy cost of manufacture. An automobile is a complex piece of hardware, and every part of it comes into being through a process that starts at an assortment of mines, oil wells and the like, and proceeds through refineries, factories, warehouses, and assembly plants, linked together by long supply chains. All this costs energy. Working out the exact energy cost per car would be an unmanageably huge project because it would involve tracking the energy used to produce and distribute every last screw, drop of solvent, etc., but it's safe to say that a very large portion of the total energy used in a car's lifespan is used up before the car reaches the dealer. Electric cars are just as subject to this rule as petroleum-powered ones.

Promoters of the more grandiose end of alternative energy projects—the solar power satellites and Nevada-sized algae farms that crop up so regularly when people are trying to ignore the reality of ecological limits—are particularly prone to brush aside the energy cost of manufacture with high-grade handwaving, but the same habit pervades nearly all thinking about energy these days. Factor in the energy cost of manufacture, and there's a straightforward answer to the question we've just been considering. If you really feel you have to have a car, what kind involves the smallest carbon footprint and the least overall energy use? A used one.

If you buy a used compact, let's say, instead of a new electric car, you've just salvaged the energy cost of manufacture that went into the used car, and saved all the energy that would have been spent to produce, ship, and assemble every part of the new car. You will also be producing less carbon dioxide than the share of a smokestack that would be needed to power an e-car. Since it's a used compact, furthermore, you won't be tempted to drive it all over the place to show everyone how ecologically conscious you are; in fact, you may be embarrassed enough to leave it in your driveway when you don't actually need it, thus saving another chunk of energy. Finally, of course, the price difference between a brand new Nissan Leaf and a used compact will buy you a solar water heating system, installation included, with enough left over to completely weatherize an average American home. It's a win-win situation for everything but your ego.

That said, of course, a car of any kind is rarely the best answer to all your transportation needs, and it may not be the best answer to any. I've never owned a car. That was one of the decisions I made back in the days of the appropriate tech movement, and it's one I've never felt a need to reconsider. When I need to go few miles from where I live, I walk; if it's more than that, but within the range of local transit, I catch a bus; if I need to travel to another city, I catch a train. I know far too many people who can't imagine themselves doing any of these things; even when they recognize the huge burden that the private auto puts on their own lives and the planet, the thought of doing without a car, even now and then, makes them quail.

Especially but not only in America, the car has been loaded down with so much in the way of powerful cultural fantasies and emotional drives that it's almost impossible to talk about it in purely practical terms. I dislike cars, and not just on principle—chalk it up to a family habit of long, pointless Sunday drives with the smoke from my father's cheap cigarettes pooling like a miasma in the back seat—but I've still felt, while catching a ride with friends to some Druid gathering, the lure of the open highway that plays so huge a role in America's collective psyche.

That's a major theme in our national character that I suspect many people elsewhere in the world simply don't get. The vast majority of white Americans are descended from people who turned their backs on the Old World to chase the dream of a better life on the other side of the ocean. That pattern of seeking a new life elsewhere has repeated far more often than not with each generation. One of the many factors that make white Americans so clueless about nonwhite Americans, in turn, is that that experience isn't shared with the other peoples of this nation. For us, that first journey beyond limitations has always defined the American experience, but for African Americans, their encounter with this continent was a bitter exile into bondage; for the Hispanic population of the United States, the defining experience was dispossession, and for the first inhabitants of this continent, it was not merely dispossession but very nearly annihilation. A road leading into the far distance means something very different to the descendants of pioneers on the Oregon Trail than it does to the descendants of those who survived the Trail of Tears.

Still, even among white Americans, the dream of freedom somewhere on the far side of the horizon could have expressed itself in many different ways. It so happens that nowadays, at least, it almost always expresses itself through the automobile. This is why Americans cling to their cars with such frantic intensity, and why Republican politicians—always a better barometer of the American mass psyche than their Democrat rivals—so reflexively treat any alternative to the private car as a threat to America's freedom. On any rational level, of course, that's the most vacuous sort of hogwash, but on a nonrational level—on the level of collective passions and mass fantasies where most human motivation takes shape—it's a potent reality.

Yet those passions and fantasies are facing a catastrophic collision with an even more powerful reality. The private automobile is the ultimate poster child for the age of cheap abundant energy, hopelessly unsustainable without immense inputs of highly concentrated energy and vast and ongoing investments in infrastructure. The freedom of the open road, in other words, was never much more than an illusion. Behind it lies the car culture's total dependence on government subsidies to build the roads and gargantuan corporations to build the cars and provide the gasoline to run them. As cheap abundant energy becomes a thing of the past, and those inputs become first unaffordable and then unavailable at any price, detaching the passions and the fantasies from what is, after all, just one means of transport will be an important step.

There are plenty of ways to start doing that. Walking, bicycling, and public transit of various kinds are all options, and there are doubtless others—though the grandiose plans for billion-dollar transportation projects popular in the high-tech end of the environmental scene are a waste of everyone's time; the resources needed to build such things will not be available in a time of contraction. If your personal passion for green wizardry focuses on a basement workshop and plans for some new transportation technology, by all means pursue it—but remember to plan for the real world, subject to the cycles and limits we've already discussed.

Exercise for Lesson 30

Run an errand or take a trip, using a mode of transport different from the one you would normally use for that purpose. If at all possible, make it a mode that uses less energy than the one you would otherwise use. Consider repeating the experience.

RESOURCES FOR PART THREE

There were at least as many books written on energy conservation and homescale energy production during the heyday of appropriate tech as there were about gardening and homescale food raising. Too little useful work has been done since that time, though. Because they require a head-on confrontation with the expectations of limitless abundance that pervade our culture, conservation and homescale energy were consigned to oblivion with the coming of the Reagan years. It's rare to find anyone even in the greenest of green circles who is willing to discuss them these days, and until recently, new books on this subject were fairly rare. Thus, you may have a hard time finding good modern editions of the classics or a good selection of new publications on the same themes at your local bookstore.

As with previous lists, the list that follows makes no pretense to completeness. It simply includes those books from the appropriate tech movement I have used and found valuable.

Butti, Ken, and John Perlin, *A Golden Thread: 2500 Years of Solar Architecture and Technology* (New York: Van Nostrand Reinhold, 1980).
　　The standard history of solar energy, essential for anyone interested in the field. The best way to skip reinventing the wheel is to see what's already been tried, what worked, and what didn't.
Campbell, Stu, with Doug Taff, *Build Your Own Solar Water Heater* (Charlotte, VT: Garden Way, 1978).
　　You'll need serious skill with plumbing and general construction to use this book for its obvious purpose, but it's also one of the best guides to the technology even for those who plan on having a manufactured system installed by a professional.
Daniels, Farrington, *Direct Use of the Sun's Energy* (New Haven, CT: Yale University Press, 1964).
　　Everybody in the appropriate tech scene who was interested in solar energy had a copy of this on the bookshelf, usually the 1971 paperback reprint by Ballantine Books. It remains among the most comprehensive studies of solar energy available; readers interested in solar thermoelectric energy will want to read Chapter 15 with care.
Fisher, Rick, and Bill Yanda, *The Food and Heat Producing Solar Greenhouse* (Santa Fe, NM: John Muir Publications, 1976).
　　Already included in the list of resources for Part Two, this classic source belongs here as well. Not to be missed.
Gay, Larry, *The Complete Book of Heating with Wood* (Charlotte, VT: Garden Way, 1974).
——, *The Complete Book of Insulating* (Brattleboro, VT: Stephen Greene, 1980).
　　Two good hands-on guides to producing heat and keeping it in the house.

Halacy, Beth, and Dan Halacy, *The Solar Cookery Book* (Culver City, CA: Peace Press, 1978).

A classic book with detailed plans and tasty recipes, this was reprinted in 1992 as *Cooking With The Sun*; by any title, it's worth having.

Kirschner, Heidi, *Fireless Cookery* (Seattle, WA: Madrona Press, 1981).

The only really good book on fireless cookers I know of, and long out of print. Some small press could do itself and the world a favor by tracking down the copyright holders and bringing out a new edition.

Langdon, William K., *Movable Insulation* (Emmaus, PA: Rodale Press, 1980).

A thorough guide to insulated window coverings, this also covers ways to insulate solar greenhouses and solar water heaters.

Mazria, Ed, *The Passive Solar Energy Book* (Emmaus, PA: Rodale Press, 1979).

The standard reference for passive solar heating, packed with tables and technical data hard to find anywhere else. If you want to explore passive solar, this is your guide.

McGuigan, Dermot, *Harnessing the Wind for Home Energy* (Charlotte, VT: Garden Way Books, 1978).

A solid introduction to homescale wind technology.

Merrill, Richard, and Tom Gage, *Energy Primer* (New York: Delta Books, 1978).

An excellent introduction to alternative energy resources, this became a standard text in appropriate tech circles.

Shurcliff, William, *New Inventions in Low-Cost Solar Heating* (Brick House Publishing, 1979).

——, *Thermal Shutters and Shades* (Brick House Publishing, 1977).

Shurcliff was among the most innovative figures in the solar energy end of appropriate tech. The first book goes beyond the standard approaches to explore dozens of novel ways to harvest heat from the sun; the second, a comprehensive survey of insulated window covering, contains more than a hundred designs.

Steadman, Philip, *Energy, Environment and Building* (New York: Cambridge University Press, 1978).

Perhaps the best overall survey of building-related energy issues to come out of the appropriate tech movement.

US Department of Housing and Urban Development, *In The Bank…Or Up The Chimney?* (Radnor, PA: Chilton Books, 1976).

There were dozens of books, pamphlets, and mimeographed handouts on energy conservation produced during the 1970s energy crises. This one is among the best, a simple and practical guide to home energy conservation.

Watson, Donald, *Designing and Building a Solar House* (Charlotte, VT: Garden Way, 1977).

A good, thorough introduction to the options for building a sun-heated home from the ground up.

PART FOUR

Whole Systems

LESSON 31
Putting It Together

At this point, it may be helpful to glance back over the ground we've covered in the last thirty lessons. From one perspective, those lessons have simply offered a grab-bag of potentially useful methods for producing and preserving some of your own food and some of your own energy on a scale small enough that an individual, a family, or a very small group could easily put them to work. In the future that is taking shape around us right now, methods like the ones we've covered will almost certainly offer better options than continued reliance on gargantuan technostructures utterly dependent on a limitless supply of increasingly limited resources.

The hope of providing better options of this kind was a central motivation of the appropriate tech movement of the 1970s and early 1980s. Still, for a great many people who were involved in appropriate tech back in the day, the goals of the movement weren't limited to the provision of helpful strategies for an age of crisis. Beyond that necessary work lay the hope of constructing, at least in outline, ways of living on the Earth that would be humane and fulfilling as well as ecologically sustainable.

These hopes rested on a tradition of social criticism that has been forgotten and buried every bit as thoroughly as appropriate tech itself. From the Great Depression right up through the coming of the Reagan era, a sequence of brilliant social thinkers—Lewis Mumford, Norman Brown, E. F. Schumacher, Theodore Roszak, and Christopher Lasch, to name only a few—subjected industrial society to close analysis and showed just how destructive it was to every human value, very much including the happiness that the cheerleaders of the industrial system insisted it was supposed to provide. Central to the thinking of all these authors was the recognition that a life spent frantically attempting to satisfy manufactured cravings for consumer products, at the expense of more authentic human needs, does not lead to fulfillment. The fact that these same lifestyles also endanger the health of our planet and the long-term prospects of our species is simply one more layer of bitter icing on an already unwelcoming cake.

Such ideas are hardly popular these days. Still, they provide a crucial part of the framework for practicing the green wizardry of appropriate tech, and attempting to integrate the practices of the last thirty lessons into your life without at least a little attention to that broader picture makes things a good deal more difficult than they have to be.

Here's an example. I routinely field emails from people who are seriously troubled about the future. They see themselves as trapped in a system that's already started to go to bits around them, but they lack the money and other resources they would need to weather the approaching crash. A good many of them are living in apartments with nowhere to garden and few options for energy retrofits, and they quite reasonably worry about what's going to happen when access to energy becomes intermittent, food prices spike, and what now counts as a comfortable urban lifestyle begins the long downhill skid into the shantytown existence many Americans will encounter in the decades ahead. They want to know what options I can suggest for them.

I do have a suggestion to offer, though it's not one that many of them are eager to take. The most important thing people in that situation can do is to get off the consumer merry-go-round and stop spending money on the dubious conveniences and even more dubious entertainments that eat such a large portion of all but the poorest Americans' incomes these days. The goal of that strategy is to bring expenditures well below income, so that the money left over can be used to get out of the current, unsustainable situation.

Most Americans can cut expenses by anything up to a third by giving up the energy- and money-wasting habits of the consumer economy. That may involve moving to a smaller apartment with lower rent, fewer amenities, and a bus line nearby; it may involve not buying the new computer every two years, the plasma screen TV, and the other expensive toys so many people think they have to have; it may involve learning to cook, eat, and enjoy rice and beans for dinner instead of picking up meals at the deli; it will likely involve plenty of other steps of the same kind. The payoff is that you get the extra money you need to learn the skills that will make sense in a deindustrial economy; then, you can save up a down payment for a fixer-upper house with good solar exposure, a backyard well suited for an organic garden, and a basement where you can get to work learning to brew good beer. For many people, using less now is the entrance ticket to a better future.

There's another side to these preparatory steps, because it's impossible to downshift in a blink from a modern American lifestyle, with all its comforts and privileges, to the close-to-subsistence lifestyle most of us will be leading in the middle future. There's no good to be gained by following the lead of those old-fashioned survivalists whose idea of being ready to feed themselves, once the rubble stops bouncing, was a nitrogen-packed tin of garden seeds, a random assortment of tools, and a manual on how to garden, which they read halfway through on a slow afternoon ten years ago. Those who adopted that approach have been very lucky that their doomsteads have never had to function as anything more serious than deer camps, because if they'd tried to feed themselves that way, death by starvation would have been the inevitable result. Growing food in an intensive organic garden is a skilled craft requiring several years of hard and careful work to master, and if you hope to rely on it for even a small part of your food, you need to get through the steep part of the learning curve as soon as possible.

The same thing is true of most of the other skills that are needed to live comfortably in hard times. If you don't know how to do them, you're going to make a lot of mistakes, which means you'll suffer a great deal more than you have to. The sooner you start that learning curve, the easier the curve will be, because you'll still have the resources to pick up the pieces when your early efforts fall flat. It's entirely possible, for example, to live through summers south of the Mason-Dixon line without air conditioning—people did it for a very long time before air conditioners were first marketed in the boom times following the Second World War—but it's not simply a matter of gritting your teeth and sweating. It requires certain skills and, for most recently built houses, certain modifications. If the thermometer hits three digits when you haven't yet installed the attic fan or figured out how to open a couple of windows at the right angle to catch the breeze and keep heat from building up, you could be risking heatstroke. Starting the learning curve now provides a margin of safety you'll be glad to have.

Most current talk about the impact of peak oil assumes that the end of the industrial age is a nice, cleanly marked point located conveniently off somewhere in the future. That's a potentially lethal oversimplification. Those Americans who have run out of their 99 weeks of unemployment checks and become members of the new class of economic nonpersons have already been pushed out the exit doors of industrial society; for them, the end of the industrial age has arrived. That same eventuality could show up on any of our doorsteps with 99 weeks of warning, and quite possibly less. If that happens to you, will you be better prepared to meet it if you've cut your expenses, cleared your debts, mastered the art of getting by with less, and learned the skills and bought the tools for a backup profession or two, or will you be better off if you've been spending everything you earn, and then some, in standard American middle class style? You tell me.

All this amounts to variations on a common theme, which is that the rules governing life in a stagnant or contracting economy are precisely the opposite of the rules governing life in an expanding one. In the growth economy of the recent past, it usually made sense to spend money freely and gamble that you could always get more, because the sheer fact of continued economic growth meant that more often than not, you would be right. With the end of economic growth, the principle once made famous by Wilkins Micawber, the amiable moneylender of Charles Dickens' novel *David Copperfield*—"annual income twenty pounds, annual expenditure nineteen pounds nineteen and six, result happiness; annual income twenty pounds, annual expenditure twenty pounds ought and six, result misery"—once again comes into force.

I don't have anything so elegant as Micawber's principle to offer. What I suggest, rather, is an acronym—LESS—that stands for "Less Energy, Stuff, and Stimulation." In outline, that's the strategy I'd like to propose for those who want to weave green wizardry into a broader way of life. Just as it's a lot easier to heat a house with solar power when you've already taken care of insulation and weatherstripping, it's a lot easier to live a life in an age of decline when you've made sure your life isn't leaking energy and other resources from every available orifice. That's what the LESS strategy is meant to do; think of it as a way of weatherstripping your life.

The last part of the acronym, "stimulation," may seem surprising, but it's a crucial part of the recipe. For the last thirty years and more, Americans have been pushing their nervous systems into continual overload with various kinds of stimulation. A mind that's constantly flooded with noise from television, video games, or what have you, is a mind that never has the time or space to think its own thoughts. In a nation that's trying not to notice that it's sold its own grandchildren down the river, that's probably the point of the exercise. Be that as it may, recovering the ability to think one's own thoughts, to clear one's mind of media-driven chatter, manufactured imagery, and all the other thoughtstopping clutter we use to numb ourselves to the increasingly unwelcome realities of life in a failing civilization, is an indispensable tool for surviving the challenges ahead.

"Stuff" may seem a little less puzzling, but getting out from under the tyranny of excess ownership may be every bit as challenging for many Americans as shaking off the habit of stimulating the mind into a state not far removed from coma. As far as I know, ours is the only civilization in history in which building and managing storage facilities for personal possessions has become the basis for a significant economic sector. It's a critical issue, though, because our passion for what I've elsewhere termed "prosthetic technologies"—machines, that is, that are designed to do things that human beings are perfectly able to do for themselves—has built up habits of dependence that could easily, and literally, prove to be fatal if they're not broken before demand destruction puts the machines and the power needed to run them out of reach. In an expanding civilization, your success is marked by what you have; in a declining one, your chances of survival may well be measured by what you can readily do without.

"Energy," finally, may be the most obvious factor in the equation, but some of its aspects are far from obvious to most Americans today. A very large fraction of the energy that props up the American lifestyle, for example, gets used to manufacture, package, ship, retail, power, maintain, and dispose of the heap of consumer goods that people in this country commonly mistake for having a life. Another very large fraction, as just suggested, goes into technologies meant to keep human bodies and minds from doing things they're perfectly able to do, and, as often as not, become unhealthy if they're not allowed to do. For every watt-hour that can be saved by direct methods, there's more than one—very often, many more than one—that can be saved by indirect methods such as buying used goods from local sources rather than new items from chain stores with intercontinental supply chains, or leaving the latest round of flashy entertainment devices to collect cobwebs in a big box store while you do something less futile with your time.

Still, the basic concept should be easy enough to grasp. The habit of living beyond our means is as much an individual problem as a collective one, and it's a significant factor keeping many people stuck in a set of lifestyles that are as unsatisfactory as they are unsustainable. Freeing up the money, the time, and the resources to make the shift to a more sustainable way of life needs to be high on the agenda of anyone who's seriously planning to deal with the cascading crises of the decades ahead of us, and using LESS may be the single most important and accessible tool for doing that.

Exercise for Lesson 31

The exercise for this lesson may already have occurred to you as you read the paragraphs above. Choose one way to use less energy, one way to have less stuff, and one way to get less stimulation in the course of your daily round. Try them for a week, and see what you learn by doing so. If the changes seem beneficial to you, consider making them permanent and/or making more changes of the same kind.

LESSON 32
The New Alchemy Option

One of the elegant things about the LESS strategy, and for that matter about appropriate tech as a whole, is that neither one locks those who adopt them into a single lifestyle. Back in the heyday of the appropriate tech movement, in fact, there were at least three distinct ways that the basic toolkit tended to be put to work, each with its own esthetic and its own typical lifestyle choices. To the best of my recollection, nobody gave them names at the time, and that's a gap that needs filling. For our purposes, they can be called the New Alchemy, Down Home Funk, and Retrofit modes of green wizardry.

New Alchemy, the subject of this lesson, takes its name from one of the most consistently innovative of the appropriate-tech nonprofits, the New Alchemy Institute. These were the people who designed an integrated bioshelter—part solar greenhouse, part fish farm, part homestead—that fed its occupants and got all its energy needs using sun, wind, and a little bit of wood heat on the coldest winter nights, and they built it in the far-from-genial climate of Cape Cod and made it work. Then they built another one on Prince Edward Island, where the weather is even worse, and made it work there, too. The New Alchemy Institute is the reason you've heard of tilapia; that tasty, nutritious and quick-maturing fish was the species they chose for their pioneering work in aquaculture. These and other projects made the arrival of each issue of *Journal of the New Alchemists* a welcome event in appropriate-tech circles.

The keynote of the Institute's work, and of the broader style of green wizardry I've named after it, was the application of technological innovation to the needs of an energy and resource-conserving lifestyle. There was a huge amount of this going on during the appropriate tech era. From wind turbines and solar power projects through novel conservation strategies to fully integrated projects like the bioshelter on Cape Cod, technological experimentation was an ongoing theme all through the history of appropriate tech, pursued enthusiastically by nonprofits such as the New Alchemy Institute as well as small businesses such as ecological designer Steve Baer's Zomeworks, which spent the entire era dazzling the competition with ingenious new ways to collect, store, and use solar energy.

The New Alchemists and their peers hoped that they were inventing the tools and insights that would move humanity into a sustainable future. Thus it's no surprise that a New Alchemy or Zomeworks building would fit right into any of the eco-themed science fiction movies of the decade. Even today, there's something eerily futuristic about the few buildings of this kind that have survived; it's just that they come from a future that doesn't happen to be the one toward which we're headed.

Even so, a strong case can be made for dusting off the pioneering work that was done along these lines back in the appropriate tech era, and an even stronger case presents itself for taking the central principles that governed these efforts and putting them to use with the options available today. While the pace of overall technological progress is often overstated, a few fields relevant to appropriate tech have undergone what deserves to be called revolutionary change since the early 1980s: advances in materials science and information processing technology, to cite only two of the most important of these, really have made it possible to do new things, and not—as so much so-called progress does—the same old things in a newer, flashier, and less sustainable way. If this is the kind of appropriate tech that appeals to you, there's much that can be done.

Still, there are certain common pitfalls in the way. One uncomfortably precise example of those pitfalls can be found in the work of one of the few appropriate tech mavens to remain a public figure right up to the present, the durable Amory Lovins. Lovins and the organization he heads, the Rocky Mountain Institute, stayed respectable during and after the Reagan counterrevolution by discarding what I've called the LESS strategy, and insisting that industrial society in its present form can be powered indefinitely by renewable sources. Arguing for that claim requires a degree of optimism verging on myopia, and that requirement is one that Lovins is apparently more than willing to accept. One notorious example is the 2000 interview in which Lovins insisted that hybrid and fuel cell cars would account for between half and two thirds of the cars on the road in the United States by 2010.[19]

Of course, he was wrong. Hybrid cars accounted for maybe five percent of the US automobile fleet that year, and you can still look through every automobile showroom in North America for a car powered by fuel cells and not find one. What needs to be understood here, though, is that there was never any reason to think he would be right. Hybrid cars may cost less to operate, but they're much more expensive to build than ordinary cars; fuel cell cars, while they could probably have been made for a more competitive price, could only compete in any other way if somebody had invested the trillions of dollars in infrastructure to provide them with their hydrogen fuel. In both cases, economics made it impossible for either kind of car to account for more than a token fraction of the US car fleet by 2010, and their chances of being much more popular by 2020, or 2030, or any subsequent year are not much better.

The specific economic factors that made Lovins' claim so much empty air are less important than the general point, which is that airy optimism about technologies that haven't yet gotten off the drawing board is not a useful response to an imminent crisis in the real world. This is a point worth keeping in mind, because airy optimism about technologies that haven't yet gotten off the drawing board is flying thick and fast just now, especially but not only in the peak oil scene. Mention how industrial society is in deep trouble as a result of dependence on rapidly depleting fossil fuels, and you can count on a flurry of claims that some vast energy project that hasn't yet reached the commercial-prototype stage—fusion reactors, algal biodiesel, or what have you—will inevitably show up in time and save the day.

I hope it will not come as a surprise if I point out that nothing of the kind is going to happen. This is partly a function of economic factors parallel to those that have kept fuel cell cars from fulfilling Lovins' prophecy—the cost of building out the new technologies and of providing them with the infrastructure they need in the teeth of competition from existing energy sources and competing alternative technologies alike—but there's another factor as well, one that unfolds from the logic of funds and flows discussed way back in Lesson 5.

Here's how it works. Building out any alternative energy system on a scale large enough to matter would require huge inputs of real wealth—not currency, which can be manufactured at will by central banks, but energy, materials, knowledge, and labor, which are a good deal harder to conjure up out of twinkle dust. As non-renewable resources deplete, the amount of real wealth that has to be put into extracting them goes up; as pollutants of various kinds build up in the biosphere, the amount of real wealth that has to be put into mitigating their impacts goes up, too. To scrape together enough real wealth to cover the cost of an alternative energy project massive enough to make a difference would simply pile another burden onto an already overloaded economy, and, if pursued with enough misplaced enthusiasm, it could conceivably become the trigger that brings the whole thing crashing down.

Even the most technocentric New Alchemy, then, needs to find responses to the crisis of our time that don't ignore the economic impacts of that crisis. For this reason, the vague speculations about untested technology that fill so many discussions about our future these days needs to take a back seat to practical techniques that have been tested in the real world and can be deployed without massive investments of scarce resources or other forms of wealth. That may seem dowdy and uninteresting to the tech-minded. To borrow a familiar metaphor, however, if your ship has already hit its iceberg and is taking on water, passing out life jackets and launching lifeboats is far less innovative and exciting than sitting around talking about some brilliantly creative new way to rescue people from a sinking boat, but it's a good deal more likely to save lives.

Are there new technologies worth exploring, and innovations worth pursuing, in the face of that troubling reality? Of course. Plenty of innovative work has been done with homescale solar and wind energy, for example, but there's no reason to think that all the low-hanging fruit has been gathered there; nearly all the funding has gone into large-scale systems for decades now, and the scientific and technical developments of the last thirty years have yet to be put to work in these fields. Some of the best work back in the day focused on creative ways to conserve energy; that's still a goal, and there are likely to be any number of clever options that haven't been tried yet. Homescale food production using more specialized tools and techniques than the shovels, hoes, and wheelbarrows of the ordinary organic gardener is another field that could use more work. There are many more possibilities besides these, waiting for basement inventors and backyard innovators to do something new and interesting—and if it comes out looking like something from an old science fiction movie, that's all the more reason to go ahead with it.

Exercise for Lesson 32

The exercises for this and the next two lessons are work for the imagination, rather than the more prosaic mental faculties you've used in previous exercises. Your task here is to imagine yourself taking up what I've called the New Alchemy approach to green wizardry, an approach centered on innovation and the development of new techniques and technologies, with all the excitements, frustrations, and technical troubles such a choice involves. Does that concept excite you, or does it leave you cold? Only you can say.

⟞ LESSON 33 ⟝
The Down Home Funk Option

The technological innovations and experiments that came out of New Alchemy Institute, Zomeworks, and other foundations and companies on the high-tech end of the appropriate tech scene represented only one of the core trajectories of the movement.

Another took the opposite approach. Instead of trying to invent the technologies of the postindustrial future, this second option focused on reviving the technologies of the preindustrial past.

It's hard to think of anything more offensive to the conventional wisdom of industrial society, then or now, but there was and is a solid argument for this approach. Whatever else can be said for or against them, the technologies used in previous eras have two advantages that most of the alternatives can't claim: they have been proven to work in the absence of abundant fossil fuels, and nobody has to spend time we may not have figuring out how to make them work. Here and there, in the less fashionable corners of the country, a good many of them are still in use, or have been used recently enough that it's possible to find people who still know how to work them. As a salute to the cultural context of many of these legacy technologies and practices, I've termed this mode of green wizardry the Down Home Funk option.

What sort of things went into that option? To a significant extent, it depended on regional factors, since, in the absence of cheap abundant energy, the toolkit that works in the hills of North Carolina is by no means the same as the one that works on the plains of Iowa. Still, certain things were commonplace throughout that part of the appropriate tech scene: wood heat, hand tools, traditional methods of home food production and storage, and the whole range of old-fashioned handicrafts. Look up the way that people in your corner of the world used to live before the electrical grid and chain stores reached them: in a nutshell, that's Down Home Funk.

During the heyday of appropriate tech, Down Home Funk had a huge constituency because of a third advantage shared by most of the older technologies: they're cheap. Having evolved in an age when today's abundance of concentrated energy was nowhere to be found, and local and national economies were accordingly smaller, the tools and trappings of the Down Home Funk lifestyle cost a great deal less—in money, but also in resources, labor, and other less abstract measures of wealth—than the things that replaced them. In the seventies, when a troubled economy made saving money a high priority for many people, older and cheaper ways of doing things had a definite appeal. That appeal had even more force among appropriate tech geeks, who normally had the thinnest of shoestring budgets to devote to their projects.

The downside of Down Home Funk, in turn, was precisely measured by one of its main advantages: it's a way of living based on older tools and methods that, compared to the lifestyles of most people in today's industrial America, looks impoverished. In a certain, narrowly materialistic sense, it *is* impoverished: Down Home Funk embraces the LESS strategy. In no uncertain terms those who pursue that approach end up using a great deal less energy, and a great deal fewer of the products of energy, than the average American does today. Still, it's the appearance of poverty, not the fact, that's the problem.

We Americans like to pretend that we're all rugged individualists, but this is actually a very conformist and status-conscious society; even those who insist that they're rebelling against that society are generally careful to follow whatever style of rebellion is trendy among people in their social class. That was the downfall of Down Home Funk. Once it stopped being trendy, a great many people who had briefly embraced it turned pale and bolted back to a lifestyle that didn't make them look poor in the eyes of their neighbors and friends.

The cultural hangover from that burst of status panic includes the notion, surprisingly common all through American society these days, that living the Down Home Funk way is a nightmare experience composed entirely of deprivation and backbreaking labor. This is popular nonsense, but it's still nonsense. It's entirely possible to be happy, comfortable, and productive while living an old-fashioned, low-energy lifestyle. Relearning the skills that make this possible is certainly a valid approach to the crisis of our time. I spent several years in the early 1980s living that way. Though I ended up transitioning from Down Home Funk to what I'll shortly describe as the Retrofit model, for reasons that included getting engaged and deciding to pursue a writing career, I made the shift with a definite sense of regret.

Thus, there's a wry amusement to be had in the common dismissal of the Down Home Funk scene—with its communes, tie-dyed skirts, and back-to-the-land ethos—as a fad of the seventies. When it comes right down to it, after all, today's high-consumption lifestyle is itself a fad, right up there with hula hoops and swallowing live goldfish. Thirty years ago, the thought that people would voluntarily put themselves at the beck and call of anybody who wanted to contact them, at all hours of the day and night, would have inspired a mix of horror and hilarity. Thirty years from now, those who now can't imagine being offline for more than a few hours at a stretch will look back on their current habits with much the same embarrassment that you see in today's fifty-something Republicans when they remember their long-haired, pot-smoking youth.

Now, of course, there's an extent to which the Down Home Funk end of the appropriate tech scene was a fad, or at least had faddish dimensions. There were people who devoted themselves to various kinds of conspicuous non-consumption, who treated some particular austerity as a litmus test while neglecting broader principles, and so on. Those dubious habits can be found in every social movement, and it's a safe bet that if the LESS strategy catches on at all, it will promptly attract a fringe of hangers-on who are in it purely to impress people. When you live in a conformist and status-conscious society, that comes with the territory.

Still, in the future looming up ahead of us, even the most faddish approach to Down Home Funk has its advantages. The ongoing crisis in the global economy, as it ripples down to the individual level, is making it increasingly hard to find the resources necessary for the kind of innovative experimentation that New Alchemy requires. It is also increasingly hard to justify gambling available resources on projects that, like all innovations, will inevitably have a high failure rate. In such a setting, tools and techniques that are already tested and ready to use have a definite value. If it takes a fad to get them into circulation, then that's what it takes.

I mention this possibility because the fad in question already seems to be taking shape. As Kurt B. Reighley pointed out in a lively recent book, *The United States of Americana*, one of the major American cultural trends of recent years has been the revival of old-fashioned traditions along a spectrum that extends all the way from heirloom tomatoes to bump-and-grind burlesque shows.[20] The popularity of historical reenactment groups of various kinds, many of them home to eager practitioners of a dizzying assortment of long-neglected skills, is another marker of the same trend, as are not-quite-historical phenomena such as the steampunk scene, with its ingenious reinventions of the Victorian technological esthetic. All these could easily help feed the return of Down Home Funk as a fashion trend in the near future.

However silly its surface forms might turn out to be, and however short a time it lasts, such a revival might be a very good thing to adopt and even to encourage. Even if it only lasts for a decade or two, that may be enough to help a lot of people weather the immediate impact of the crisis to come. Whatever fashions emerge in its wake, though, it's safe to say that today's fad for frantic consumption won't be among them—for the simple reason that the resources that would make that fad possible are going away forever. Whatever fads and fashions spring up in the aftermath of the approaching crisis will have to make do with a much smaller resource base.

A fashion for simpler, old-fashioned practices may be temporary, in other words, but the habits thus established may well endure. We are all, every one of us, going to have to get by with less energy and fewer of the products of energy; we are all going to have to do things for ourselves that we've come to assume, often unthinkingly, that machines powered by cheap abundant energy will always do for us; we are all going to have to accept a great deal more in the way of discomfort and inconvenience than we do today. That has to be accepted, or nothing constructive is going to get done.

One effective way to minimize those discomforts and inconveniences is to learn and practice the skills needed to get by with much less energy than we do today. The Down Home Funk option offers one way to do that, the New Alchemy option offers another. The differences between them are partly esthetic, partly—as already mentioned—that the Down Home Funk option relies on proven options, while New Alchemy is experimental and thus a gamble against uncertain odds. Both of them have a place in any renaissance of appropriate tech, the former to provide the solid foundation of reliable approaches, the latter to search out new options that might bring significant benefits as we proceed further down the road to the deindustrial future.

(Exercise for Lesson 33)

Your assignment for this lesson parallels the one for the previous lesson. Imagine yourself returning to older, lower-tech ways of doing things in your daily life and making those new habits permanent. Does that thought fascinate you or make you uncomfortable? Again, only you can say.

✈ LESSON 34 ✈
The Retrofit Option

The New Alchemy and Down Home Funk versions of appropriate tech both have much to recommend them as applications of the LESS strategy, and more generally as ways of approaching green wizardry in practice. Both had large and lively constituencies back in the 1970s and early 1980s. Still, there was a third option at the time, and it was by most measures the largest and liveliest of all. It drew its popularity from a simple fact: neither New Alchemy nor

Down Home Funk is necessarily well suited to the environments in which most people in the industrial world live right now.

The would-be practitioner of New Alchemy is worst off where this issue is concerned. Nearly all the designs for really energy-efficient houses turned out by the designers and architects of that end of the appropriate tech scene, for example, presuppose that you're starting with a vacant lot, often with specific requirements— for instance, a gentle south-facing slope for maximum solar gain and convenience in putting an earth berm against the house's northern side, plus no high buildings or trees within a hundred yards or so to maximize the efficiency of the wind turbine, and so on. That's great if you happen to have such a piece of property or can afford to go shopping for one; it's also great if you can afford to build a house, or have one built for you, from the bare ground up. Plenty of people don't have that option, however.

Down Home Funk, as already mentioned, is easier to manage on a budget. The tools and techniques that go into it were devised in an age when most people had a lot less wealth available, but it also presents its own difficulties. Most people in the industrial world nowadays don't live in an old-fashioned farmhouse or even its urban equivalent, a city row house with a good-sized garden in back. Moreover, the kind of houses and apartments that have been popular with builders and developers for the last half century or so lack most of the useful features that anybody buying or renting a home in 1890 would have expected as a matter of course.

Does this mean that people who can't afford to build an earth-sheltered, solar-heated, wind-powered New Alchemy Institute-style bioshelter, or a nice hundred-year-old home with a big working kitchen, a sizeable pantry, a root cellar, and room for a vegetable garden and dwarf fruit trees out back, are out of luck? Not at all, and that's where the third option comes in. This is the option of taking existing resources of every kind, from the built environment right down to the contents of the nearest junk heap, and repurposing them to make them serve goals that were not what their designers, manufacturers, or sellers had in mind.

Because the habit of retrofitting existing houses with good insulation, storm windows, and solar energy collectors was one of the most visible expressions of this approach during the heyday of appropriate tech, we'll call it the Retrofit option. Still, readers who are familiar with my previous books on peak oil know another name for the same approach: it's the basic style of what I've called the salvage economy—the form of economy we'll have when the energy surplus necessary to support an industrial society goes away and the most readily accessible resources are to be found in the buildings, products, and refuse of the departed industrial system.[21]

Whether or not that turns out to be the wave of the future, it was a viable option in the days of the appropriate tech movement, and it's still a viable option today. Most of today's houses, for example, can fairly easily be retrofitted for energy efficiency—that's much of what the previous lessons on energy conservation were about, after all—and, given a decent southern exposure or a lack of tall trees nearby, can equally well be retrofitted to collect some of its occupants' energy needs from sun, wind, or both. Most of today's backyards can readily be repurposed to make room for vegetable gardens, henhouses, rabbit hutches, and the like. Most lifestyles can be reworked to make more use of bicycles, shoe leather, public transit, and other transportation modes and much less use of the energy and resource-wasting car culture, and the list goes on.

One of the most important steps in embracing the Retrofit option, in turn, is one that terrifies most Americans: meeting their needs with used items instead of new ones. This follows the logic discussed a few lessons ago in terms of automobiles: in most cases, the best way to save energy and resources is to buy something used. It's only our society's class prejudices that make this seem difficult. Very few people, for example, actually need a new computer. I've never owned one. I need a computer to make my living as a writer—publishers these days require book manuscripts to be submitted electronically—but I get my computers used, free or at a tiny fraction of their original price, and run them until they drop. One result is that I've salvaged the energy and resources that went into the old computer, rather than burdening the economy with the energy and resource costs of a new one; another is that I've kept a small but measurable amount of toxic e-waste out of the waste stream; still another, of course, is that I save money that can then be directed to other purposes, such as insulation and garden tools.

Most Americans buy most of the things they use new, and dump perfectly useful items into the trash; the more conscientious package up their used items and donate them to thrift stores, which is at least a step in the right direction. As a society, we've been able to afford this fixation and its many costs because we've been surfing a tidal wave of cheap abundant fossil fuel energy. That state of affairs is rapidly coming to an end, and the Retrofit option is one good way of responding to that shift. Right now, while the tidal wave of cheap energy has not yet receded very far, the beachscape of industrial society is still littered with the kind of high-quality salvage our descendants will dream of finding. The only thing that has to be overcome in order to access most of it is the bit of currently fashionable arrogance that relegates used goods to the poor.

Those of my readers who decide that the advantages of voluntary poverty are worth pursuing are going to have to confront that attitude in themselves, if they haven't done so already. Like all supposedly classless societies, America makes up for its lack of formal caste barriers by raising class prejudice to a fine art; the cheap shots at small town America so common among the urban middle classes who dominate today's green scene are an expression of that, and so is the peer pressure that keeps most Americans from doing the sensible thing and buying cheap and sturdy used products in place of increasingly overpriced and slipshod new ones.

We are all going to be poor in the decades and centuries to come. Yes, I'm including today's rich in that; the stark folly that leads today's privileged classes to think they can prosper while gutting the society that guarantees them their wealth and status has been displayed many times before in history, and it will bring about the usual consequences in due time. Voluntarily embracing poverty in advance may seem like a strange tactic to take, at a time when a great many people will be clinging to every scrap of claim to the fading wealth of the industrial age, but it has certain important advantages. First, it offers a chance to get competent at getting by on less before sheer necessity forces the issue; second, it sidesteps the rising spiral of struggle that's waiting for all those who commit themselves to holding on to an industrial-age standard of living; third, as I've already pointed out, buying cheap used items frees up money that can then be applied to something more useful.

Those who opt out of current social habits and embrace a Retrofit lifestyle will have to deal with a certain amount of social pressure, but they will also have certain advantages not available to their more conformist neighbors. The additional money in the budget already mentioned is only one of these. Another will be the chance to learn skills that could well become income sources in the not-too-distant future; anything that involves taking the leftovers of industrial civilization and turning them into something that people need or want, will likely be among the major growth industries of the next century or two. The ground floor is open to entrepreneurs right now.

Exercise for Lesson 34

Your exercise for this lesson complements those for the two previous lessons. Imagine yourself pursuing a Retrofit lifestyle, buying used goods instead of new ones, modifying your existing home for energy conservation, and exploring the possibility of a new career in some salvage-based trade. Does that prospect intrigue you or appall you? Here again, only you can say.

The Way of Dissensus

The last three lessons have another point to communicate beyond the explicit one of discussing three once-popular ways to integrate appropriate tech into your life. The fact that there are three of them, and not just one, raises an issue of some importance, one that needs to be addressed as we finish this survey of the green wizardry of the appropriate tech movement.

It's a common but counterproductive habit of thought these days to think that every problem has a single solution and that the best way to reach that solution is for everyone to work together to pursue it. If the environmental sciences have any lesson to teach us, by contrast, it's that every problem has many possible solutions, and the best way to find and use as many of those solutions as possible is for everyone to "work apart," so to speak, heading off in as many disparate directions as possible, so the largest possible number of options will be tried.

This is the logic of dissensus, a concept introduced by postmodern theorist Ewa Ziarek in a 2001 book on ethical theory.[22] Dissensus is exactly what it sounds like, the opposite of consensus. More precisely, it is the principled avoidance of consensus, and it has its value when consensus, for one reason or another, is either impossible or a bad idea—when, for example, irreducible differences make it impossible to find any common ground on the points that matter, or when settling on any common decision would be premature.

This last situation, I suggest, is a fair description of where we stand now. There's an interesting dichotomy in our knowledge of the future: history can give us a fair idea of the type of events that we will encounter, but neither it nor anything else can give us the details. Looking at what's happened to previous civilizations that overshot the limits of their resource base, for example, it's not hard to recognize the parallels and predict the onset of the familiar process of decline and fall. What we cannot know in advance is just how these general patterns will play out in space and time.

Lacking those details, a consensus plan is not a good idea. If you knew today, let's say, that the region where you live was going to have much less rain in the future, you would make one set of choices; if you knew that the same region was going to have much more rain in the future, you would make another, and so on. If you knew that a million refugees from climate change will be coming through your town, your plans would be very different from the ones you would make if you knew that your town would be far from the migration routes.

Since these things can't be known in advance, though, whatever consensus you reach has a very real chance of being exactly the wrong choice. This is where dissensus comes to the rescue. In a situation of uncertainty, encouraging people to pursue different and even conflicting options increases the likelihood that somebody will happen on the right answer.

A dissensus approach is valuable right now for another reason. We are far past the point at which an organized, society-wide program to deal with the crisis of industrial civilization is possible—as the Hirsch report pointed out back in 2005, that had to start twenty years before the peak of petroleum production,[23] which puts that hope more than a quarter century into the realm of might-have-beens. Even if the option still existed, the political will simply isn't there. That means that ad hoc responses are probably the best options we've got. Focusing on those possibilities that can be done on a shoestring, and maximizing the total number of these that get tested in the immediate future, is thus a crucial strategy right now. Even if most of those efforts fail, this approach will yield the largest number of useful ways to manage crises as they occur and make some degree of recovery possible later on.

This logic has at least one implication that probably won't sit well with many of my readers: that people should be encouraged to pursue projects that, according to the best current evidence, have little or no apparent chance of succeeding. That's a necessary consequence of a dissensus-based approach; as Charles Fort used to say, "It is by thinking things that schoolboys know better than to think that discoveries are made." [24] The one caveat that has to be added, though, is that anyone advocating any such project actually needs to be doing something about it.

The Bussard fusion reactor makes a good example. It's a modest variation on the Farnsworth fusor, a laboratory curiosity dating from the 1950s that uses the same technology as an old-fashioned picture tube to fuse very small numbers of hydrogen nuclei into helium. Decades of study and experimentation have shown precisely no evidence that it can be made to produce more energy than it uses up. For several years now, the Bussard reactor has nonetheless been cited over and over again by people on internet forums as a reason not to worry about peak oil.

There are a modest number of people—according to recent media reports, around a dozen—who have built Bussard-style devices in their basements and achieved nuclear fusion, confirmed by neutron detectors. None of them are producing an energy gain, or anything close to it, at this point, and there's no reason to think they ever will. Still, I have the utmost respect for these people; they're putting their money, as well as their time and energy, where their mouths are. If there's any chance that Bussard was right, this is how we'll find out.

The flip side of this is simple enough: if people want to insist that the Bussard reactor is going to save us all, the appropriate response is, "Are you building one?" If they are, well and good; if they're chipping in as much as they can afford to help cover the expenses of someone who's building one, that will pass; otherwise, they're simply engaging in handwaving and may be disregarded by the rest of us.

The same principle can be applied to any other proposed response to the crisis of industrial society. If it's viable as a basement-workshop project, then anybody who intends to claim that it's a solution to the crisis of our time had better be building one, or helping to pay the bills of someone else who's building one. If it's too large, complex, and expensive for individual entrepreneurs to pursue, it's probably going to be too large, complex, and expensive for a civilization caught in the jaws of fossil fuel depletion, climate instability, and economic unraveling. There are some exceptions—the rebuilding of America's rail system comes to mind—but in that case there are still ways to contribute, at least to the extent of the cost of a round-trip ticket whenever you need to travel.

This is one of the reasons why I've limited my focus in this book to things that I do myself, or have done in the past and am gearing up to do in the future as soon as funds and time permit. The kind of SUV environmentalism that waxes rhapsodic about all the things everybody else ought to do for the environment, while doing few or none of them, is not a viable response to the crisis of our time. I'm willing to open my mouth about energy conservation and organic gardening, appropriate tech and antique tech, doing without and doing with less, because these are things that I do myself. I'd hardly offer myself as any kind of paragon of virtue—there's much more that I could be doing—but I'm not going to advocate what I'm not willing to do.

On the other hand, if somebody is out there working on some proposed response, they ought to be given the benefit of the doubt, not to mention the respect due to anybody who's trying to live up to their aspirations. I would extend that rule very far. The biodynamic agriculture devised most of a century ago by Austrian mystic Rudolf Steiner, for example, combines quite a few very pragmatic steps—Steiner's books are where modern organic gardening got the idea of raised beds, for example—with such things as planting by astrological influences, that most people reject out of hand these days. I know people who use Steiner's methods, and they seem to get good results; if planting by the stars and mixing weird herbal concoctions into their compost helps them grow crops and keep people fed during the times to come, that's what matters.[25]

This is the point of dissensus. It's impossible to know in advance what set of tools and skills will be best suited to squeak past the mess taking shape around us as a civilization and technology predicated on limitless expansion runs headlong into the hard limits of a finite planet. Green wizardry is only one option, and even within green wizardry there needs to be plenty of room for different paths, new inventions, local traditions, and a good helping of outright eccentricity.

Evolution depends on variation; that's what provides the raw material that gets sorted out by natural selection. There's no shortage of natural selection piling up for us in the not-too-distant future, that's for sure, but it's up to us to provide the variation. That's part of what's behind my recommendation that aspiring green wizards haunt used book stores and go digging for obscure pieces of information, and it's also part of the reason why I've focused on 1970s appropriate tech, even though there are many other options out there. The logic here is straightforward: a good number of green wizards who didn't grow up in that decade are bound to get irritated by the deliberately annoying seventies-centric approach I offer here. If that irritation sends them looking for other options, all the better!

Exercise for Lesson 35

For this lesson's exercise, choose something improbable that you think might offer a possible response to any of the aspects of the crisis of industrial society, and get to work on it. If that involves piecing together a Farnsworth fusor in the basement, good; if it involves learning how to plant by the Moon, good; whatever it involves, if it appeals to you, get on with it. Don't leave it to someone else; do it yourself, because that's the only way it's going to happen.

Passion can't be legislated, and the sort of passion that led, for example, Gregor Mendel to spend years crossing pea plants to tease out the secrets of heredity is one of the things the world needs most right now. At worst, you'll be able to draw a line under an unhelpful approach so that resources can go elsewhere; at best, you may just provide the world with some small but valuable piece of the puzzle of survival. If we're to reach the future's further shores with any of the more useful legacies of the last three centuries intact, that willingness to take personal responsibility for making things happen is essential.

LESSON 36
The Long View

All humor aside, this book's focus on seventies-era organic gardening, energy conservation, and the like is more than an exercise in encouraging readers to try something else. It is not simply a matter of nostalgia for a time when America seemed to be on the brink of taking its future seriously, or for that matter a recognition that we don't have a lot of time left and would be wise to concentrate on options that have already had the bugs worked out. Rather, by some combination of prudence, prescience, and plain dumb luck, the toolkits of ecotechnic options pieced together by the backyard farmers, basement inventors, shoestring-budget nonprofits and local government initiatives of that time happen to be very well suited to one of the dominant features of the future ahead of us.

To understand how this works, it's necessary to look at some of the least welcoming features of that future, and that in turn requires a glance over a way of thinking about the future that I explored in detail in my book, *The Ecotechnic Future*.[26] In that book, I divided the next few centuries into four overlapping phases, labeled according to the modes of economic production that predominate during each one. The first of these, the one in which most of us grew up and to which nearly all current political, economic and social thought is attuned, is *abundance industrialism*. This is the phase in which the supply of goods and services available to people in the world's industrial nations increases with each passing year.

Familiar and normal as it seems to most people in the industrial world today, this phase is ending. With the arrival of hard geological limits to fossil fuel production, we enter the second phase, *scarcity industrialism*. This is the phase in which the supply of goods and services available to people in the industrial nations begins to contract, year after year. According to mainstream economic doctrines, that can't happen, which may be why we've become so good at ignoring it.

Few people thus notice that most of what's for sale in supermarkets is a little smaller and a little more shoddily made with every passing year, while the price stays level or creeps upward; few people talk about the disappearance of scores of once-common services—try to get a perfectly good shoe with a worn heel repaired in most American cities nowadays—or think about the way that municipal services always seem to contract while the cost always seems to expand. All these changes are part of the same process: the emergence of scarcity industrialism. That phase ends when the level of goods and services being produced drops below the level needed to support any kind of industrial system at all.

After that come *salvage societies*, which no longer have the energy per capita needed for industrial modes of production and concentrate on extracting value from the leftovers of the industrial past. Later still, the salvage era winds down as the salvage runs out, and new societies based on renewable resources take their place. In a fit of optimism, I called this latter phase the ecotechnic era, and suggested that it could see relatively advanced technology supported on a truly sustainable basis. I still think that's possible, though it's going to take quite a bit of work now, and even more in the centuries to come, to make it likely.

Still, it's the age of scarcity industrialism that deserves close attention right now, since the industrial world is on the trajectory that leads there. Some core features of developed societies in the age of scarcity industrialism aren't too difficult to predict, partly because equivalent things have happened in past eras of overshoot and decline, and partly because some nations right now are much further along the trajectory than others and provide a useful glimpse ahead.

The role of social conflict in such times is fairly predictable. In an age of abundance, the easiest way to deal with social conflict is to buy off the disaffected. That's how industrial societies over the last century came to provide welfare for the poor, mortgage guarantees and college grants for the middle class, subsidies for farmers, tax breaks for businesses. Name a group that's had enough political savvy to organize and raise a ruckus, and you can just as quickly name the arrangements by which they were paid off to minimize the risk of disruptions to the system. That was politically feasible in an expanding economy; even when the shares of the existing pie were grossly unequal, the fact that everyone could have at least a little more each year made those with smaller slices willing to work with the system in order to get their cut.

In an age of scarcity, that easy option no longer exists, and social conflicts heat up rapidly. That's the unmentioned subtext for much of what's going on in politics on both sides of the Atlantic just now. The middle class, which shrugged and turned its collective back when the working classes of Europe and America were thrown to the sharks thirty years ago, is now discovering to its horror that it's next on the list.

This is happening because the rentier class, the fraction of industrial society that makes its living from investments rather than salaries, is struggling to maintain its prosperity at everyone else's expense. (The middle classes did exactly the same thing when they had the chance—ask any impoverished working class family in Pittsburgh or Glasgow—so sympathy cards sent their way may be misplaced.) The gutting of social safety nets, the slashing of salaries

and benefits, and the impoverishment of millions of previously af-
fluent people are part of that process, and they all lead to a rising
spiral of social conflict that may well push a good many nations
into crisis or collapse.

It might, in theory, be possible for a country to extract itself
from a descent into social conflict along these lines. The chance
that the industrial world will manage such a last-minute save,
though, is minimal. Here in the United States, we're already seeing
basic government services slashed to a degree unequalled in the
industrial world; a social safety net that was already an embarrassment for a developed nation is headed for the chopping block; the
machinery of government in state houses and Congress alike is
jamming up as pressure groups of every kind launch increasingly
frenzied efforts to cling to wealth and influence at everyone else's
expense.

It's not a pretty picture, and it's unlikely to get any more attrac-
tive any time soon. Still, it's important to understand why societies
in decline so often plunge into this self-defeating spiral. One of
the major problems faced by a society in decline is their almost
universal unwillingness to deal with the fact that they are indeed in
decline. There's a straightforward solution to the many difficulties
spawned by this unwillingness, but the solution—accepting that
the glory days of the past are over, and that the new and unwelcome
reality of contraction and limitation will be around for the foreseeable future—is accepted only after every other imaginable response
or nonresponse is tried out, and found wanting. An overabundance
of absurd beliefs, grandiose projects, and violent political passions
are standard means of avoiding that one necessary step, and there
are plenty of examples all around us.

There's at least one way out of the trap, however. The declining years of a rich and powerful society resemble nothing so much
as a game of musical chairs in which all the chairs will sooner or
later be taken away. What's the winning strategy in a game in which

everyone loses sooner or later? That's a simpler question than it sounds: the way to win is not to play the game.

And that, in turn, is what we've been talking about in this book: how not to play the game.

The struggles of the age of scarcity industrialism will focus with increasing bitterness and intensity on access to the remaining benefits of industrial society—above all, the cheap abundant energy that powers automobiles and planes, keeps wall sockets supplied with electricity, brings foodstuffs and consumer goods from around the world, and provides the context and the income for jobs in the increasingly overlapping spheres of business, government, and the military. The struggles for these things, if historical equivalents are anything to go by, will focus on certain geographical and social battlefields and have less concentrated effects everywhere else. Those who turn their backs on the things being fought over, and distance themselves from the battlefields, have a very good chance of staying clear of the resulting difficulties.

That's an important part of what the green wizard toolkit is meant to do. Those who can produce at least some of their own food, concentrating on high-nutrient foods such as vegetables and small livestock, are no longer wholly dependent on an increasingly unsafe industrial food system. Those who insulate and weatherize, and provide some of the energy they need from homescale sources, will be sheltered from many impacts of the decline of the electrical grid. Those who learn to provide as many as possible of the things they need by their own efforts, instead of relying on global supply chains fed from depleting resource stocks, will be able to stand aside as what's left of the global economy circles the drain. Those who do these things, teach these things to neighbors and friends, and help build local networks of mutual exchange and support, will be creating the frameworks for the next stage of the future—the stage of salvage societies—within the crumbling skeleton of the old industrial order.

This, again, is what the skills outlined in this book are meant to do. To a very real extent, that's what they were designed to do, since a great many people who were involved in the appropriate tech movement back in the day expected industrial society to falter and start coming apart around them, leaving the task of rebuilding a viable human society to those individuals and local groups willing to take on that challenge. That sense is well worth reviving now, as the unraveling that was predicted in the 1970s becomes an inescapable fact.

Exercise for Lesson 36

The exercise for this lesson consists of imagining the rest of your life—however long that is likely to be—against the background of a future of gradual economic contraction, social turmoil, and technological regression. Your goal is to envision ways in which you and those you care about can lead creative, humane, and meaningful lives in such a time. This often takes a major effort of imagination for people nowadays, but the effort is worth making; it's those who can reimagine their lives in a way that doesn't rely on the crutch of faith in perpetual progress who will be best prepared to accomplish things worth doing in the years to come.

RESOURCES FOR PART FOUR

Books that attempted to weave the available appropriate tech options together into a vision of whole systems and possible futures were extremely common in the 1970s and early 1980s. Many of them were vague, ungrounded in practical experience, and bubbling over with the sort of enthusiastic optimism that's common in writers who have never done the things they're advising. Fortunately, there were also efforts along these same lines that avoided these pitfalls and offer much of value to today's green wizards. The books listed below are among those that have influenced me most.

Alexander, Christopher, Sara Ishikawa, and Murray Silverstein,
   A Pattern Language (New York: Oxford University Press, 1977).
   A brilliant and still unsurpassed study in how architecture and
   town planning might serve human needs, deeply informed by
   an appropriate tech esthetic.

Baer, Steve, *Sunspots* (Seattle, WA: Madrona Publishers, 1979).
   Baer was the visionary engineer behind Zomeworks, the most
   innovative of the appropriate tech startups of the 1970s. This
   anthology of essays and stories on the theme of solar energy is
   a good introduction to his style of ecological alchemy.

Boyle, Godfrey, and Peter Harper, eds., *Radical Technology* (New
   York: Pantheon Press, 1976).
   This British anthology was among the most popular guides to
   appropriate tech as the foundation for a new society. Primarily
   Retrofit in its vision, but elements of New Alchemy and Down
   Home Funk also fit in here and there.

Brand, Stewart, ed., *The Next Whole Earth Catalog* (New York:
   Rand McNally, 1980).
   "Access to tools" was its subtitle, but it served as an access
   point to the culture and ideas of the entire appropriate tech
   movement. The earlier editions—*The Whole Earth Catalog,
   The Last Whole Earth Catalog*, and *The Whole Earth Epilog*—
   are also well worth reading; editions after 1980 abandoned
   appropriate tech for a mix of overhyped computer geekery and
   Left Coast cultural fashions, and have little of value to offer.

Editors of *Rain, Rainbook* (New York: Schocken Books, 1977).

Editors of *Rain, Stepping Stones* (New York: Schocken Books, 1978).
   Rain was among the best of the appropriate tech magazines,
   and these two anthologies collect some of the best ideas and
   essays from its pages. A reliable source of ideas and insights for
   green wizards.

Farallones Institute and Sim Van der Ryn, Helga Olkowski, Bill
   Olkowski, Tom Javits, *The Integral Urban House* (Gabriola
   Island BC, Canada: New Society Publishers, 2008).

The bible of Retrofit, based on the experiences of an appropriate-tech nonprofit that took an ordinary urban house and turned it into a self-sufficient, energy-conserving, food-producing, waste-recycling green marvel. This book covers all the details, too.

Freed, Dolly, *Possum Living* (New York: Universe Books, 1978). The appropriate tech scene of the 1970s had close ties to the live-on-almost-nothing ethos that emerged from the 1960s counterculture, and this lively guide to getting by on next to no money was highly and deservedly popular back in the day.

Johnson, Jerry Mack, *Down Home Ways* (New York: Times Books, 1978).
The Down Home Funk version of appropriate tech generated a great deal of interest in old-fashioned ways of getting by, and this in turn produced a bumper crop of books by people who had, or claimed to have, knowledge of those ways. Jerry Mack Johnson's books were among the best contributions, and this is a particularly good collection.

Leckie, Jim, Gil Masters, Harry Whitehouse, and Lily Young, *Other Homes and Garbage* (San Francisco: Sierra Club Books, 1975).
Despite the title, this classic appropriate tech manual is squarely in the New Alchemy camp, with detailed tables and other information resources for a remarkable range of conservation, energy generation, and food production schemes on a homescale basis.

Papanek, Victor, *Design for the Real World* (New York: Pantheon Press, 1971).
A brilliant introduction to design principles, this was deservedly popular among appropriate tech mavens back in the day.

Soleri, Paolo, *Arcology: The City in the Image of Man* (Cambridge, MA: MIT Press, 1969).
Way off at the upper end of New Alchemy, located somewhere in the undefined middle ground between visionary brilliance

and pretentious grandiosity, Solari's science of arcology was an attempt to design whole cities as single, ecologically sustainable structures. Unless you have access to limitless funding, this is anything but a practical guide, but as a source of inspiration for less gargantuan ideas, it definitely has a place.

Todd, John, and Nancy Jack Todd, *Tomorrow is Our Permanent Address* (New York: Harper & Row, 1980).

Todd, Nancy Jack, ed., *The Book of the New Alchemists* (New York: E. P. Dutton, 1977).

The New Alchemy Institute was one of the brightest stars in the firmament of appropriate tech nonprofits, and its work remains a vital resource for principles and practical tips today. These two books are the best introduction to their work, and to the style of appropriate tech I've named New Alchemy after them.

Wigginton, Eliot, ed., *The Foxfire Book* (Garden City, NY: Anchor Books, 1972).

This book and its sequels were the bibles of Down Home Funk back in the day, as well as a testament to an extraordinary educational experiment—a magazine of local folklore, *Foxfire*, staffed and produced entirely by high school students in one of the more impoverished corners of the Appalachians. You can still get more practical information on old-fashioned subsistence farming skills here than anywhere else.

Wing, Charles, *From the Walls In* (Boston: Little, Brown, 1979).

The most widely used handbook, back in the day, on the fine art of retrofitting a house for energy conservation and comfortable living, and thus a solid introduction to the Retrofit approach.

Why It Matters

I've suggested already that the toolkit of practical methods and useful concepts assembled by the appropriate technology movement provides one of the best options we have as the age of cheap abundant energy winds down and a very different world of scarcity begins to take shape around us. Propose that concept to most people in the various green movements active nowadays, however, and odds are you'll hear a flurry of reasons why none of the old appropriate tech is applicable today. It's a fascinating reaction, and it has roots that are not necessarily apparent at first glance. If you sit down over a couple of beers with someone who reacts that way, and encourage them to keep talking, the core issue that will come out sooner or later has nothing to do with appropriate technology as such. It links into an almost visceral discomfort with the cultural setting from which appropriate tech emerged, the lively and idealistic alternative culture of the 1970s.

It's easy to make fun of the excesses and eccentricities of the era: the air of well-scrubbed, fresh-faced innocence, say, that was so assiduously cultivated by the exact equivalents of those who now cultivate an equally artificial aura of hardbitten practicality or sullen despair. Still, the remarkable creativity of the appropriate tech movement and other now-extinct cultural initiatives of that time suggests that those John Denver fans, with their granny glasses and dogs sporting brightly colored bandanas in place of collars, had something going for them that today's supposedly more sophisticated culture has not been able to match. The shift from the one to the other set of cultural themes may have more to do with that difference than is often recognized, and we need to explore that possibility to finish making sense of the way that the green wizardry explored in this book might contribute to a more livable future.

That is to say, we need to talk about the roots of the contemporary cult of nihilism.

I don't think that last phrase is too extreme a description. For the last few decades, in America and much of the rest of the industrial world, it's been hugely fashionable to believe, or at least to affect to believe, the cynical notions that all ideals are frauds or delusions, that those who try to live up to them are either posturing liars or simple-minded fools, and that we might as well enjoy ugliness because all beauty is by definition fake. Watching this week's idols dragged down to the lowest common denominator by yet another wretched scandal has become America's most popular spectator sport.

Meanwhile, and crucially, the notion that people might face a challenge, any challenge, by rising to the occasion, gets dismissed out of hand by pundits, politicians, and ordinary people alike—if the possibility is mentioned at all. This wasn't always the case. As this nation and the industrial world as a whole lurches blindly toward a set of challenges right up there with anything in the last five thousand years or so of recorded history, it bears asking why a rallying of will and potential that would have been an obvious part of a response to crisis fifty years ago is so unthinkable now.

It's useful, in making sense of this cultural shift, to remember that there are at least two kinds of cynicism. There's the kind, variously weary, amused, hurt, or icily dangerous, that comes naturally to those who have too often seen others betray their ideals. Then there's the other kind—sullen, jeering, brittle, and defensive—that comes just as naturally to those who betray their own ideals, making them lash out angrily whenever anything too reminiscent of that betrayal flicks them on the raw. It's the latter kind, I'm convinced, that shapes the mood of America today. The disquieting sounds that murmur through the crawlspaces of our collective imagination, waking us abruptly at night, are the echoes of a profoundly troubled national conscience.

For another measure of the same troubled conscience, think of the extraordinary reach of conspiracy theories of all kinds through American culture. These days, if you hear people talking about any of the problems or predicaments that beset our society, it's usually a safe bet that the conversation will promptly focus on some group of people whose monstrous wickedness is allegedly the cause of it all. Democrats talk that way about Republicans, and Republicans about Democrats, while those who have abandoned the grinning corpse of America's once-vital political culture have their own colorfully stocked rogues' galleries of alleged villains to offer.

If you want to see how much of this fixation on hunting for scapegoats unfolds from an uneasy conscience, try suggesting in public that ordinary people might bear some modest degree of responsibility for the unwelcome trends of the last few decades. The shrill tones in which most Americans will insist that all the blame lies elsewhere makes it clear just how sensitive a nerve has been touched. What Carl Jung called "projecting the shadow" has become a potent political reality in America, but you don't need a degree in Jungian psychology to realize that people who spend their lives pointing fingers at other people are trying to paste a villain's mask on the rest of the world in order to avoid seeing it when they look in the mirror.

A third measure? Consider the contemporary American obsession with apocalyptic fantasies. Back of all the gaudy claims of history's end currently on display—the Rapture, the Singularity, and all the rest of it—is a frantic insistence that we don't have to live with the consequences of our collective actions. That's the common thread that connects the seeming optimism of the claim that Jesus or the Space Brothers or superintelligent computers will fix things, on the one hand, with the seeming pessimism of the claims that we're all about to be wiped out by solar flares or asteroid bombardment or the evil plans of the Illuminati. Either way, the world that our choices have made is not the world we have to inhabit. Either way, it's not our responsibility to fix what we've broken, either because someone else is going to fix it or because it's all going to be blown to smithereens shortly by something that, please note, is never our fault.

All three of these factors have deep roots in American history, but it's not too hard to identify the point in time when they moved in from the fringes to dominate the collective imagination—and that lands us once again in the wake of the seventies, the years when a society that had previously idolized John-Boy Walton and John Denver suddenly started idolizing Gordon Gekko and self-proclaimed "material girl" Madonna instead.

A glance back over the history of the second half of the twentieth century will help put that shift into context. The aftermath of the Second World War left the United States abruptly filling the position of global hegemon previously held by Great Britain. In the aftermath of Hitler's defeat, Americans believed they had a permanent lease on the moral high ground. Mixed motives and the pressures of expediency had their usual effect, though, and as the cognitive dissonance built up, it became increasingly hard for Americans to pretend that all the atrocities and abuses of the Cold War era belonged to the other side.

Those pressures reached critical mass in the early 1970s. The Pentagon's epic incompetence in the Vietnam war and the blatant illegality and corruption of the Nixon administration sparked a backlash that, for once, reached right up into the corridors of power. In the wake of the resulting explosions, American troops came home from Southeast Asia, Nixon was forced out of office, and a quarter century of dubious and often illegal policies unexpectedly saw the light of day. All this took place during the runup to the US bicentennial, and the contrast between admittedly idealized notions of the 1770s and the awkward realities of the 1970s forced many Americans to notice the gap between what they had become and what they claimed to be.

These cultural shifts were occurring just as America's own oilfields reached their all-time peak production; the coming of America's own encounter with peak oil threw into question a generation of easy assumptions of perpetual national prosperity. There were still plenty of people alive who vividly recalled the Great Depression and the austerity of the war years, and for that reason could get their minds around the idea that the postwar boom might be a temporary event, or even a corner into which the United States had backed itself. Many Americans, across a wide range of social and political positions, embraced the possibility that a prudent regard for resource limits might be a valid approach to economic and political questions, and that resource conservation and a shift toward less extravagant ways of living might be the best available options over the long term. An even broader spectrum of Americans came to believe, at least for a time, that something crucial to their nation's meaning and value had gotten lost in the rush to global empire, but might still be recovered in time to matter.

It's popular nowadays to forget that this happened, or to insist with varying degrees of cynicism that the moment of awareness couldn't have lasted. Maybe that's so, but I wonder how much of that insistence comes from the same uneasy conscience that drives so much of today's fashionable nihilism. Americans came together during the long ordeal that began with the stock market crash of 1929 and wound its way through the shadows of depression and war until 1945. A similar effort over a similar time scale would have been more than adequate to the task of launching America into the transition to an ecotechnic future. Back then, the United States still had abundant coal, oil, and natural gas reserves, not to mention a great many other resources; annual consumption of energy and resources was far below what it later became, and, though a great many factories were shuttered in the sharp recessions of the 1970s, there were still millions of capable laborers who could have been put to work retooling the economy for a new and frugal age.

The steps necessary to make that transition were discussed in any number of periodicals, some of them surprisingly mainstream. The consensus was that the United States would have to step back from its self-appointed role as global policeman, pass on a fair share of the cost of deterring the Soviet Union to its comparatively more prosperous allies in western Europe and the west Pacific, and accept a less expansive notion of its own national interests. Government subsidies for nuclear power and other nonrenewable energy sources would need to be phased out, and the money, along with savings from a less gargantuan military, could be shifted into grants for conservation, renewable energy retrofits, and research programs aimed at repositioning American industry to lead the world in green energy technologies.

Changes in tax policy, zoning regulations, and building codes would have reshaped the built environment to decrease energy use, while funds formerly wasted on highways would go instead to build high-speed rail between urban cores and rapid transit systems that would make commuting by car all but obsolete. All this would have cost plenty, and it would have required Americans to tighten their belts and accept a diminished standard of living, including some formal or informal rationing, at least for a time. Down the road a quarter century or so, though, a prosperous nation getting by comfortably on a fraction of its previous energy needs would be able to ignore the Middle East as an irrelevance; the US would have the lion's share of global trade in new energy technologies, high-speed rail, and a dozen other fields, while other nations burdened with high energy costs would be left scrambling to catch up.

That was the vision. Again, it's comforting to the collective conscience of today's America to insist that it couldn't have happened, but "comforting" is rarely a synonym for "true." Myself, I think that there was a very real chance of making it happen. The uncomfortable silence that falls when anyone brings up the subject of conservation in most circles today is one of the reasons I've come to that belief. When people set aside an obvious impossibility, they don't remain brittle and angry about it for decades afterward; it's only when the road not taken was a real option, and the goal at the end of it noticeably better than the endpoint looming up ahead, that those who chose otherwise get shrill in defense of their decision. That shrill tone is hard to miss these days, and it's grown in volume and intensity over the course of the thirty year vacation from reality the industrial world took in the aftermath of the Seventies.

The impact of that choice is perhaps easier to trace on the conservative end of the social and political spectrum. Forty years ago, for example, the Republicans had as good a record on environmental issues as the Democrats, and the idolatry of the unrestrained free market that pervades the Right these days was a fringe ideology widely, and rightly, considered suspect by most conservatives. For that matter, creationism and speculations about the imminence of the end times were consigned to the fringes by most Christians, who by and large considered them irrelevant to the task of living a life centered on the teachings of the gospel.

All these things changed in a hurry at the end of the seventies. Why? Because the attitudes that replaced them—the strident insistence that the environment doesn't matter, that the free market will solve every problem, that the world was created in 4004 BCE with as much oil, coal, and gas as God wants us to have, and that the world will end in our lifetimes so our grandchildren won't have to deal with the mess we'd otherwise be leaving them—are all attempts to brush aside the ugly fact that the choices made at the end of the seventies, and repeated by most Americans at every decision point since then, have cashed in the chance of a better future for our grandchildren, and spent the proceeds on an orgy of consumption in the present.

The squirmings of the leftward end of culture and politics are a little subtler, since the Left by and large responded to the end of the seventies by clinging to its historic ideals, while quietly shelving any real attempt to do anything about them. It's the discomfort with this response that leads so many people on the Left to insist angrily that they've done all they can reasonably be expected to do about the environment, even as they pursue lifestyles that are impossible to distinguish from those of their conservative neighbors. This same discomfort also drives the Left's frankly delusional insistence that everyone on Earth can aspire to a middle-class lifestyle if the evil elites would simply let it happen; and the same discomfort fuels the equally, if more subtly, delusional claim that some suite of technologies currently in the vaporware stage will permit the middle classes of the world's industrial nations to have their planet and eat it too.

Look beyond the realm of partisan quarrels, and the same deeply troubled conscience appears over and over again. Consider, as one example out of many, the way that American culture transformed protecting children from a reasonable human concern to an obsessive-compulsive fixation. Raised under the frantic surveillance of helicopter moms, forbidden from playing outside or even visiting another child's home except on the basis of a prearranged and parentally approved play date, a generation of American children were held hostage by a galaxy of parental terrors that had only the most distorted relationship to reality. All of this served to distract attention from the fact that the lifestyles chosen by these same parents were condemning their children to a troubled and dangerous life in a depleted, polluted, and impoverished world.

The irony reached a dizzying intensity as tens of thousands of parents rushed out to buy SUVs to transport their children to places every previous generation was perfectly capable of reaching

by themselves on foot or on bike. It became the conventional wisdom, during the peak of the SUV craze, that the safety provided to
young passengers by these massive rolling fortresses justified their
purchase. No one wanted to deal with the fact that it was precisely
the lifestyle exemplified by the SUV that was—and remains—the
single most pressing threat to children's long-term safety and welfare.

A great many of the flailings and posturings that have defined
American culture from the eighties to the present, in other words,
unfolded from what Jean-Paul Sartre called "bad faith"—the unspoken awareness, however frantically denied or repressed, that
the things that actually mattered were not things anyone was willing to talk about, and that the solutions everyone wanted to discuss
were not actually aimed at their putative targets. The lie at the heart
of that bad faith was the desperate attempt to avoid facing the implications of the plain and utterly unwelcome fact that there is no
way to make a middle-class lifestyle sustainable.

Let's repeat that, just for the sake of emphasis: *There is no way to
make a modern middle-class lifestyle sustainable.*

That's the elephant in the living room, the thing that most of
a continent and much of the industrial world has been trying not
to see, and not to say, for all these years. The middle-class North
American lifestyle, to borrow and extend James Howard Kunstler's
useful description of suburbia, is an arrangement without a
future;[27] it's utterly dependent on the rapid exploitation of irreplaceable resources, and the longer that it's pursued, and the more
people pursue it, the worse the consequences will be for children
now living, and for a great many generations not yet born. It really
is as simple as that.

Now it's not at all hard to find books, films, websites, or speakers
making these points, but it's intriguing to watch how universally
they avoid the next logical step. What do you do if you're pursuing
a way of life that has no future? Well, apparently, you read books
denouncing that way of life, or heap praise on cultures conveniently

distant in space or time that you think had or have or will have a different way of life, or you engage in token activities intended to show that your heart really isn't in that way of life, or you vent your rage against whoever it is that you blame for your decision to keep on following that way of life, or you fixate with increasing desperation on prophecies insisting that the Rapture or the Singularity or the space brothers or somebody, anybody, will bring that way of life to an end for you so that you don't have to do it yourself.

The one thing you don't do is the one thing that actually matters: changing the way you live here and now.

That's the rock on which the sustainability movement of the seventies broke, and it's claimed plenty of victims since then. The climate change movement is a good recent example. There were plenty of reasons why the climate change movement followed the trajectory it did—from apparent unstoppability a decade ago to its dead-in-the-water status today. The ingenuousness with which climate change activists allowed their opponents to redefine the terms of the debate very nearly at will, and the movement's repeated attempts to rest its arguments on the faltering prestige of science in an age when most people are well aware that scientific opinions can be purchased to order for the cost of a modest grant, did not help the cause any.

Still, I've come to think that the Achilles' heel of the entire movement was the simple fact that none of its spokespersons showed any willingness to embrace the low-energy lifestyle they insisted the rest of the world had to adopt. Al Gore, with his sprawling air-conditioned mansion and his frequent jet trips, was the poster child for this, but he had plenty of company. It was because climate change activists so often failed to walk their talk, I suggest, that millions of Americans decided they must be making the whole thing up. In addition, the obvious eagerness of the United States and its allies to push carbon limits on every other nation—while refusing to accept them at home—convinced China among others that the global warming crusade was simply one more gimmick to prop up the crumbling edifice of American hegemony and brought the movement toward a worldwide carbon treaty to the standstill where it remains today.

The same blind spot continues to plague what's left of the climate change movement. Consider erstwhile environmentalist Stewart Brand, who used to edit *The Whole Earth Catalog*, for heaven's sake. Brand's current position is that we have to run our economy on nuclear power because burning coal is bad for the environment.[28] Now, of course, this argument, logically speaking, is right up there with insisting that shooting yourself through the head is good for your health because it prevents you from dying of a heart attack. There's a deeper irrationality here, however, and it's one that most people who had copies of *The Whole Earth Catalog* on their shelves forty years ago could have pointed out in a Sausalito minute. Switching from one complex, centralized, environmentally destructive energy system based on nonrenewable and rapidly depleting resources to another energy system that can be described in exactly the same terms, is not a useful step—especially when it would be perfectly possible to dispense with both by simply using less energy.

The concept of using less of anything is about as popular in the contemporary industrial world as garlic aioli at a convention of vampires. Nobody wants to be reminded that using less, so that our grandchildren would have enough, was the road we didn't take at the end of the seventies. Still, the road we did take was always destined to be a dead end, and, as we move deeper into the twenty-first century, the end of that road is starting to come into sight. At this point, we're faced with the prospect of using less energy, not because we choose to do so but because the energy needed to do otherwise isn't there anymore. That's the problem with living as though there's no tomorrow: tomorrow inevitably shows up anyway.

This late in the day, our remaining options are starkly limited. Most of the proposals you'll hear these days are simply variations on the theme of chasing business as usual right over the nearest cliff. Whether it's nuclear power, "Drill, Baby, Drill," ethanol, algal biodiesel, or some other kind of handwaving, the subtext to every widely touted response to our predicament is that we don't need to use less. The same thing's just as true of most of the ideologies that claim to offer a more global response to that predicament; the one common thread that unites the neoprimitivists who claim to long for a return to the hunter-gatherer life, the conspiracy theorists who spend their days in an increasingly frantic orgy of fingerpointing, and the apocalypticists who craft ever more elaborate justifications for the claim that somebody or other will change the world for us, is that each of these ideologies, and plenty others like them, function covertly as justifications to allow believers to keep on living an ordinary industrial-world lifestyle right up to the moment that it drops away from beneath their feet.

The one option that doesn't do this is the one next to nobody is willing to talk about, and that's using less.

Mention that option in public, and inevitably you'll hear a dozen different reasons why it can't help and won't matter and isn't practical anyway. Can it help? Of course it can; in a time when world crude oil production has been bouncing against a hard ceiling for most of a decade and most other energy sources are under growing strain, any decrease in the amount of energy being wasted on nonessentials makes it a little easier to keep essential services up and running. Will it matter? Of course it will; as we move into a future of hard energy constraints, the faster at least a few people get through the learning curve of conservation, appropriate tech, and simply making do with less, the easier it will be for the rest of society to follow their lead and learn from their experience—even if it only happens after all the other choices have been foreclosed. Is it practical? Of course it is; the average European gets by comfortably on less than a third of the annual energy budget of the average person in North America, and it's been my experience that most middle-class Americans can slash their energy use by a third or more in one year by a relatively simple program of home weather-izing and lifestyle changes.

I'd like to suggest, in fact, that at this point in the trajectory of industrial civilization, any proposal that doesn't focus on using LESS—"less energy, stuff, and stimulation"—simply isn't serious. It's hard to think of any dimension of our predicament that can't be bettered, often dramatically, by using less of our remaining re-sources; it's even harder to think of any project that could yield significant gains as long as people in the industrial world cling to a lifestyle that history is about to relegate to the compost bin. I'd also like to suggest that any proposal that does embrace the concept of using less should not be taken seriously until and unless the people proposing it actually use less themselves, preferably by adopting the measures they urge on others.

That's how effective movements for social change happen, after all. Individuals start them by making changes in their own lives; as the number of people making those changes grows, networks emerge to share information, resources, and encouragement; the networks become the frame of a subculture, and, as momentum builds, the subculture becomes a movement. The two movements that had the most impact on American culture in the second half of the twentieth century—feminism and Christian fundamental-ism—both emerged this way; they were started by individuals who changed their own lives first. Movements of the same era that tried to make change from the top down—again, the climate change movement is a good example—failed to achieve their ends.

This is the core concept behind the green wizardry discussed in this book. It's entirely possible for each of us to help kickstart a movement toward sustainability by using less energy and fewer natural resources in our own lives. Proven methods and mature technologies exist to help us accomplish this goal. It so happens that I learned some of those back in the late 1970s and early 1980s, when I was a geeky teenager getting underfoot in one out-of-the-way corner of the appropriate tech scene, and I've had some decades of experience applying them in my own life. That's been the basis on which I've selected the tools and techniques discussed here; for reasons already explained, I don't think it's useful to advocate things I haven't used myself.

The great barrier in the path of starting a movement the right way, beginning on the individual level, is that it requires each person who takes up the challenge to break with the conventional wisdom and do things that others aren't prepared to do. It's my hope that at least some of the people who read this book will rise to that challenge. We have a lot of work to do, and there may not be much time to get it started before conditions become a good deal more difficult than they are right now.

There doesn't need to be a sudden catastrophe for that to happen. Millions of people in the industrial world are finding themselves shut out of their familiar middle-class existence, and few if any of them will ever find a way back into it. Over the years to come, more and more people will undergo the same profoundly unwelcome shift, until what used to be a normal middle-class existence becomes a thing of the past for everybody. That's the inevitable shape of our future, because there's no way to make a middle-class lifestyle sustainable, and if something can't be made sustainable, it won't be sustained.

That doesn't mean that we're all going to move into cozy lifeboat ecovillages or any of the other green-painted Levittowns that fill so much space in so many environmental fantasies today. It means, rather, that in the decades ahead of us, something like half the US population, and smaller but still significant fractions of the population in other industrial nations, will wind up living in shantytowns without grid electricity, running water, or sewers; they will be caught up in a scramble for survival that many will inevitably lose. It means that most of the others will likely face a reduction in their standards of living to levels not too different from the one that the poorest in the industrial world experience today. It also means that the rich of that time, if they're smart, ruthless, and lucky, may be able to scrape together some of the luxuries a middle-class family can count on today, and may even be able to hold onto them for a while.

Does the picture I've painted seem unbelievable? It's simply the equivalent of saying that the world's industrial nations will undergo the usual pattern of severe economic contraction that's a normal part of the decline and fall of a civilization. What all this implies, in turn, is that those people who make the shift to a low-energy lifestyle in advance, before the sheer pressure of circumstances forces them to do so, will have options that are closed to those who cling to the unsustainable until it's dragged out of their grip. Those who downshift hard, fast, and soon, cutting their dependence on fossil fuels and the goods and services that fossil fuels make available, will have a much less difficult time paying off debts, finding the money to learn new skills, and navigating the challenging economic conditions of life in a near-bankrupt society.

These are the kinds of steps that leave people in possession of a home, a garden, a career doing something people need or want badly enough to pay for even in a depression, and other desiderata of hard economic times. These are also the kinds of steps that make it easier for people to offer help to their families, friends, and neighbors, to teach vital skills to those who are willing to learn them, and preserve precious cultural legacies through the crises of the present to they can be handed on to the future.

What we are talking about, to borrow a phrase from Henry David Thoreau, is voluntary poverty. The founders of the modern movement of "voluntary simplicity" backed away uncomfortably from the noun in Thoreau's phrase, and thereby did themselves and their movement a huge disservice. It's all too easy to turn "voluntary simplicity" into a sales pitch for yet another round of allegedly simple products at fashionably high prices. The concept of voluntary poverty does not lend itself to such evasions. The idea—Thoreau's idea, that is—is to deliberately embrace being poor in order to avoid the common fate of being possessed by your possessions.

That's a valid choice at any phase of history's wheel, but it takes on greater importance in an age when being anything but poor makes you a target for practitioners of involuntary poverty who will gladly scramble right over you as they pursue a fading vision of extravagant wealth. This is why monasticism works so well in the declining years of civilizations and the dark ages that follow them; successful monastic traditions embrace strict poverty. Having nothing to steal, they don't need to worry about thieves, and a traditional habit of choosing locations well away from the centers of wealth and power is also worth noting: Monte Cassino in Italy, the Shaolin Monastery in China, and Koyasan in Japan, where St. Benedict, Bodhidharma, and Kobo Daishi respectively founded three of the world's great monastic traditions, were all more or less in the middle of nowhere when the first simple dwellings were erected there.

What most people in today's industrial nations do not know, and have no interest in learning, is that it's possible to be poor in relative comfort. (One of the advantages of a writer's career, with its traditional slow start, is that I had ample opportunity to learn this.) The central secret of green wizardry is that one way to be poor and comfortable is to learn how to work with nature so that natural processes take care of many of your needs. The appropriate technology movement of the seventies was predisposed to develop along this path by its roots in the sixties counterculture, which (however briefly and inconsistently) embraced the ideal of voluntary poverty. Most of today's chatter about solar technology focuses on grid-tied PV systems, vast arrays of mirrors in the Nevada desert, solar satellites, and the like, which are all things that can be built only in an economy of abundance with resources to spare and thus are useless in the future we're facing. By contrast, the equivalent talk in the seventies as often as not focused on homebuilt solar water heaters, solar ovens, and other things that could be cobbled together in a basement out of salvaged materials, and thus are relevant to our time and the time ahead of us.

Whether the ideas, skills, and technologies of the old appropriate tech movement are best suited for the deindustrial world to come is ultimately for the future to decide. The crucial point now is to resurrect these things in time to get them through the next forty or fifty years of crisis, the next step down on the road to that future age. If the things we've been discussing here get pulled out of the dumpster where our society puts its unwanted heritage and are put to use by people who aren't terrified of the concept of voluntary poverty, the Benedicts, Bodhidharmas, and Kobo Daishis of the future will at least have the option of making use of the green wizardry this book has sketched out. So will a great many less exalted individuals whose lives could well be made happier and better by the application of a little ecotechnic knowledge and a few pieces of highly appropriate technology, and so, dear reader, may you.

Notes

1. Greer, John Michael, and Christopher Warnock (trans.), *Picatrix* (Iowa City, IA: Adocentyn Press, 2010).
2. Riché, Pierre, *Education and Culture in the Barbarian West* (Columbia: University of South Carolina Press, 1976).
3. von Baeyer, Hans Christian, *Maxwell's Demon* (New York: Random House, 1998) is a good nontechnical introduction to thermodynamics.
4. Hardin, Garrett, *Filters Against Folly* (New York: Penguin, 1985), pp. 30–33.
5. See, for example, Kurzweil, Ray, *The Singularity is Near* (New York: Viking, 2005).
6. Bateson, Gregory, *Mind and Nature: A Necessary Unity* (New York: Bantam, 1979), pp. 76–77.
7. Catton, William R., Jr., *Overshoot: The Ecological Basis of Revolutionary Change* (Urbana, IL: University of Illinois Press, 1980), p. 95.
8. Duhon, David, *One Circle* (Willits, CA: Ecology Action, 1985).
9. For a good introductory book on soil, see Bear, Firman, *Earth: The Stuff of Life* (Norman, OK: University of Oklahoma Press, 1962).
10. See Lewis, C. S., *The Abolition of Man* (NY: Macmillan, 1955) for a careful exploration of the roots of biophobia.
11. Kuhn, Thomas, *The Structure of Scientific Revolutions* (Chicago: University of Chicago Press, 1962).
12. Tudge, Colin, *Neanderthals, Bandits, and Farmers: The Origins of Agriculture* (New Haven, CT: Yale University Press, 1998).
13. See Tuttle, Merlin D., *America's Neighborhood Bats* (Austin: University of Texas Press, 2005).
14. A good starting place is The Xerces Society, *Attracting Native Pollinators* (North Adams, MA: Storey Publishing, 2011).
15. Taylor, Ronald L., *Butterflies In My Stomach: Insects in Human Nutrition* (Santa Barbara, CA: Woodbridge Press, 1975).
16. Buhner, Stephen Harrod, *Herbal Antibiotics* (Pownal, VT: Storey Communications, 1999).
17. Butti, Ken, and John Perlin, *A Golden Thread: 2500 Years of Solar Architecture and Technology* (Palo Alto, CA: Cheshire Books, 1980), pp. 70–72.
18. Halacy, Beth, and Dan Halacy, *Cooking With The Sun* (Lafayette, CA: Morning Sun Press, 1992), p. 41.
19. aspousa.org/2010/07/optimism-harsh-realism-and-blind-spots-years-later/, accessed December 28, 2012.

20. Reighley, Kurt B., *The United States of Americana: Backyard Chickens, Burlesque Beauties, and Handmade Bitters* (New York: Harper Perennial, 2010).

21. Greer, John Michael, *The Ecotechnic Future* (Gabriola Island, BC: New Society Publishers, 2009), pp. 70–74.

22. Ziarek, Ewa Plonowska, *An Ethics of Dissensus* (Stanford, CA: Stanford University Press, 2001).

23. Hirsch, Robert L., Roger Bezdek, and Robert Wendling, *Peaking of World Oil Production: Impacts, Mitigation, and Risk Management* (Washington, DC: US Department of Energy, 2005).

24. Fort, Charles, *Wild Talents* (New York: Cosimo, 2006), p. 317.

25. Steiner, Rudolf, *Agriculture* (London: The Bio-Dynamic Association, 1984).

26. Greer, op. cit., pp. 57–76.

27. In, for example, Kunstler, James Howard, *The Geography of Nowhere* (New York: Free Press, 1993).

28. Brand, Stewart, *Whole Earth Discipline* (New York: Viking Penguin, 2009).

Index

About the Author

JOHN MICHAEL GREER — Archdruid, historian, and one of the most influential figures in the peak oil movement, writes the widely cited weekly blog "The Archdruid Report" and has published more than twenty books on nature, spirituality and the future of industrial society. His involvement in sustainability issues dates back to the early 1980s, when he was active in the Appropriate Technology movement and became certified as a Master Conserver. He is the author of numerous titles, including *The Long Descent*, *The Ecotechnic Future* and *The Wealth of Nature*. He lives in Cumberland, MD, an old mill town in the Appalachians, with his wife Sara.

If you have enjoyed *Green Wizardry*, you might also enjoy other

BOOKS TO BUILD A NEW SOCIETY

Our books provide positive solutions for people who want to
make a difference. We specialize in:

**Sustainable Living • Green Building • Peak Oil
Renewable Energy • Environment & Economy
Natural Building & Appropriate Technology
Progressive Leadership • Resistance and Community
Educational & Parenting Resources**

For a full list of NSP's titles, please call 1-800-567-6772 *or check out our website* at:

www.newsociety.com